THE

IMPLEMENTS OF AGRICULTURE.

BY

J. ALLEN RANSOME.

LONDON:

J. RIDGWAY, 169, PICCADILLY.

MDCCCXLIII.

LONDON: R. HODSON, PRINTER, HOLBORN HILL.

This facsimile edition first published 2003

ISBN 1-903366-47-X

Old Pond Publishing
104 Valley Road
Ipswich IP1 4PA
United Kingdom

www.oldpond.com

Printed and bound in Great Britain by Biddles Ltd

PREFACE.

Of the matter contained in the following pages a considerable portion was first collected for, and formed part of, an essay for the Royal Agricultural Society of England. The work is the result of laborious observation and inquiry, which in various ways have been afforded from the circumstance of my engagement as a partner in a long established manufactory of agricultural implements. The facilities which this position has afforded me for obtaining a general knowledge of farming implements, and for becoming acquainted with the manufacture and trial of most of the machines in general use, leads me to hope that the information thus gathered, may not be without value, to those who are interested in the progress of Agricultural mechanics.

With this view, I have ventured to sketch a history of the implements in past and present use, and to introduce such comments upon their practical utility, as may possibly assist the inquirer, in arriving at a correct knowledge of their merits.

In obtaining materials for the historical part of this work, I have availed myself as much as possible of original sources of information—have explored the musty records of the patent offices—

consulted the earliest records of agriculture, and in comparing my results with those of some earlier labourers in the same field, I have pleasure in bearing testimony to the value and general accuracy of their conclusions. To the editors of the Encyclopædias of Agriculture and British Husbandry, this meed of praise is especially due. I am also indebted for much information to the works of Gray, Bailey, Williamson, Professor Low, and to the author of the article on Husbandry in the Penny Encyclopædia, whom I have ventured to quote as authorities of unquestionable value, while in the more ready and available advantage of the experience of my senior partners, I have had that assistance which gives me a just confidence that my labours as an author, although of small literary pretension, will be useful to agricultural readers; and if this object is in only a slight degree attained ample indeed will be my reward for all the pains I have taken to render the work a concise and accurate hand-book of the Implements of Agriculture.

Ipswich, 1843.

TABLE OF CONTENTS.

SECTION I.—PLOUGHS.

SECTION III.—HARROWS & SCARIFIERS.

SCARIFIERS.

SECTION IV.—ROLLERS.

SECTION V.—DRILLS.

DIBBLING MACHINES.

SECTION VI.—HORSE HOE.

SECTION VII.—RAKES.

SECTION VIII.—THRASHING MACHINE.

SECTION IX.—WINNOWING MACHINE.

SECTION X.—CHAFF ENGINES.

SECTION XI.—TURNIP CUTTERS.

SECTION XII.—MILLS.

SECTION XIII.—THE DYNAMOMETER.

SECTION XIV.—STEAM ENGINES.

THE

IMPLEMENTS OF AGRICULTURE.

SECTION I.

THE Implements which mankind have employed in the cultivation of the earth, and their gradual improvement, is a theme closely interwoven with the history of agriculture.

In tracing the gradual progress of farming implements towards their present state of perfection, it will be readily perceived how steadily, in all ages and countries, they have improved as agriculture has advanced, and how stationary they have ever remained in those countries where the science of agriculture is neglected. It would even seem that there is an intimate connection between the establishment of freedom of thought and of action, and the progress of agricultural arts and agricultural life, of all modes of life the most conducive to health, to virtue, and to enjoyment. The cultivation of the soil necessarily requires the construction of implements for the purpose; and it is gratifying to observe the progress which has been made in them in Holland, in America, and in England, and contrast the beautiful and labour-lessening implements of agriculture which these free countries possess, with those of the cultivators of Spain, of Portugal, and of Russia, or of the more degraded slaves and ryotts of the countries of the East, such as those of Palestine, and of the banks of the Ganges. These, it is more than probable, have remained unaltered, without any successful attempt at improvement, for two thousand years. Thus we find that the Israelites, instead of employing in their warm climate a thrashing machine, or even a flail, to thrash out their corn, were accustomed to turn their oxen on to the barn floor to slowly tread out the seed. And this rude mode is still the custom in Syria, and even in Portugal; and "the Moors and Arabs" says Dr. Shaw, in his Travels in Palestine, "still continue to tread out their corn in this way." "In no parts of Hindostan," observes G. W. Johnson, in his excellent Essay upon the Agriculture of India, "is the crop stacked for any time after reaping; barns are unknown; thrashing immediately succeeds the cutting of the crop. This operation is generally performed by five or six oxen, upon one of which a driver is seated, travelling abreast around a post placed in the centre of the floor. The Hindoo sacred laws expressly forbid the animals being muzzled while thus employed, but in a bad season the ryott is compelled to limit this privilege of his four-footed servant. The thrashing floor is merely a space in the field which has been levelled and beaten hard for the purpose. In Mysore the thrashing floor is made of a compound of clay, cow-dung, and water, spread over the ground, and made smooth. Some of the smaller seeds are beaten out by means of bamboo rods, but anything resembling the British flail is unknown."

And, in accordance with this neglect of labour-lessening implements, scarcely any expedients beyond the most primitive appear to have been adopted in the cultivation of the earth. Thus we find the prophet Isaiah declaring (xxxii. 20), "Blessed are they that sow beside still waters, that send forth thither the feet of the ox, and the ass." Sir John Chardin, and others, have described an indolent practice still prevalent in the Oriental countries, which explains this expression of the prophet. It seems that in planting rice, which is a crop that only flourishes in wet swampy grounds by the banks of rivers, while the earth is yet covered with water they cause it to be trodden by oxen, asses, &c., and that, after the upper portion of the ground has been thus imperfectly disturbed, they sprinkle the rice on the surface of the water.*

And if the ground is thus rudely prepared to receive the seed by the action of the feet of cattle, in a manner equally imperfect is the seed covered with the earth by these untutored cultivators. The English farmer must not expect to find in these ill-farmed and unenlightened countries any instruments even remotely resembling the compact and powerful harrows of this country: instead of these, the branch of a tree, or a few logs of wood fastened coarsely together, and dragged slowly over the surface of the very thinly and partially disturbed soil by oxen, are the only means employed to cover the seed. These instruments are thus described by G. W. Johnson: "When the plough has done its utmost on the stiff soils of Bengal, they still remain cloddy, and unfit to be seed beds. To

* And every thing relating to the preparation of the ground for the growth of corn, or its after preparation for food, appears to have been in keeping. The Israelites had formerly only hand mills to grind their corn, and these were commonly worked by female slaves.—Exodus ii. 5; Judges vi. 21; Isaiah xlvii. 1, 2; Matt. xxiv. 41.)

remedy this a still more imperfect implement than the Indian plough is employed, which is intended to produce the combined effects of the roller and the harrow. This is nothing more in form than an English ladder made of bamboo, about eighteen feet long drawn by four bullocks, and guided by two men, who, to increase its power, stand upon it as they direct, and urge on the cattle. Again and again has it to pass over the same surface, and then, as in the case of their plough, it causes a great expense of time and labour without any commensurate effect. The Indian ryotts show their consciousness of the reason that the operation of pulverizing and levelling is beneficial, by calling it *Rasbandham*, that is, *the confining of the moisture.* —*Asiatic Res.*, vol. x. p. 4.

And, in countries somewhat more civilized, the construction of agricultural implements has hardly progressed more rapidly than in the East, for even in many parts of Europe they still use ploughs of the heaviest and most ill-constructed character. Their teams, too, are equally neglected; horses, cows, asses, and even goats, are harnessed together in a most wretched manner, as was the custom, it would appear, in very primitive times in Palestine. "Thou shalt not plow with an ox and an ass together." The German farmers still use, instead of a plough, an instrument called a haken, which is exactly similar to one used by the Roman farmers. Their harrows have commonly only wooden teeth, and are worked with five horses in a very bungling manner. (*Johnson's Farm Encyclopædia*, p. 559.) And still farther north, the Muscovite harrows are formed even in a ruder way, by merely fastening together the branches of the fir tree, whose projecting, partially trimmed spurs form the teeth, while the implement they use for a plough is little more than a shapeless bundle of sticks tied together with tarred rope.

As long, in fact, as men continued to till the earth as slaves, sowing a crop they were not sure of reaping, degraded in spirit, and totally uneducated, it was in vain to expect superior implements of any kind, or any efforts, however slight, towards the improvement of agriculture. In our own island, for instance, ploughs were, during the early and dark ages of its history, rudely constructed, intolerably heavy, and of all kinds of shapes; a result which might have been reasonably anticipated, for by an old British law every ploughman was required to make his own plough. The harrows and other agricultural implements were equally ill-shaped. Drills were utterly unknown until about the sixteenth century. And when, about the year 1730, the celebrated Jethro Tull endeavoured to banish the flail from the barn, his neighbours loaded him with execrations. The tradition of the neighbourhood of Prosperous farm, near Hungerford, which Tull cultivated, still is, that he was "wicked enough to construct a machine, which, by working a set of sticks, beat out the corn without manual labour." This is the first traditionary notice of a thrashing machine with which I am acquainted. Jethro Tull, indeed, must ever be regarded as one of the earliest improvers of English agricultural implements; his ploughs, his horse-hoes, and his ingenious attempts to construct a drill machine evince a spirit of inquiry, and an advance in agricultural mechanics, which betoken at once his ability and his enthusiasm. He was far indeed before the general agricultural knowledge of his age; and if he did now and then suffer his enthusiasm to carry him too far in the conclusions at which he arrived, still the very effort to improve in such hands was sure to be attended with a measure of success; for while his exertions produced immediate good fruits, they also widely diffused a very general and well-founded suspicion that the implements of that age were not so perfect as they might be made. This led to considerable improvements, and prepared the way for still more important efforts by the next generation of implement makers, to whose merits I hope to do justice when treating of their several improvements in the implements of agriculture; and in the following Essay I shall adopt what may perhaps be fairly regarded as the natural course, beginning with the implements necessary for the preparation of the ground for the reception of the seed; then examining those adapted for the committal of that seed to the soil, for covering it with earth; for cleansing the land as the crop proceeds towards maturity; for gathering it in when ripe; and finally for thrashing out the seed and dressing or preparing it for market.

SECTION II.

THE PLOUGH.

The Plough (a name which appears to be derived from the Saxon *Plou*,) is certainly the most valuable and the most extensively employed of all agricultural implements.

The first notices of the plough are brief and slight; we find, however, that in very early times the children of Israel ploughed with two oxen (*Deut.* xxii. 10.), that their plough had a coulter and ploughshare (1 *Sam.* xiii. 20), and that they were early aware of the advantages of a winter's fallow. (*Prov.* xx. 4.) It is certain that ploughs were long since furnished with wheels; a fact which is proved by the drawings of the early Greek ploughs which have escaped to us, of which the annexed is one copy.

Hesiod (*Works and Days*, p. 50—441) advised the Greek farmers to have a spare plough, that an accident might not interrupt the work; and he also enforces the advantages of careful and skilful ploughing.

The ploughs of Rome were of the most simple form; the following engraving of one of them is inserted in this place, in order that the gradual progress of the art of

plough-making may be the more readily traced.

Rivalling these in simplicity and rudeness of form, are the never altered or improved ploughs of the Hindoos and the Chinese, from whose implements it is probable the shape of those of Rome were borrowed. This may be seen from the following sketches.

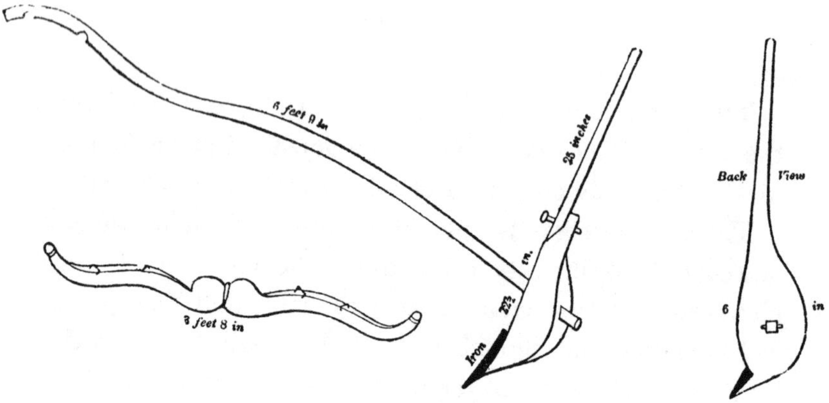

MANGALORE PLOUGH.

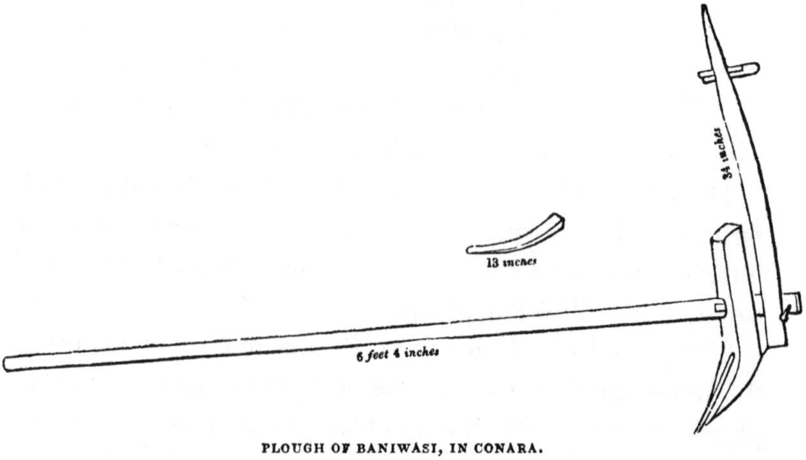

PLOUGH OF BANIWASI, IN CONARA.

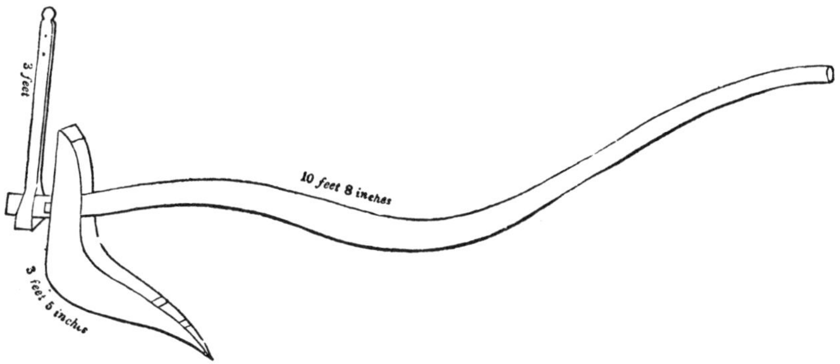

CHATAKRAL PLOUGH.

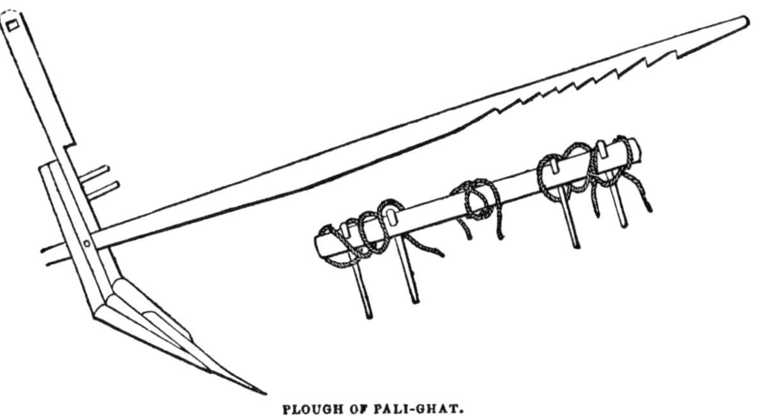

PLOUGH OF PALI-GHAT.

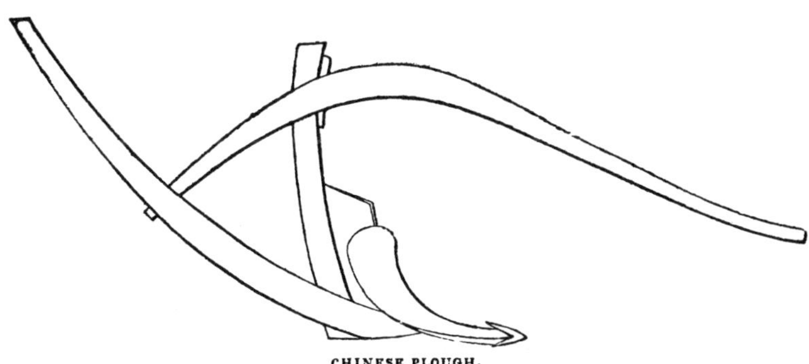

CHINESE PLOUGH.

It is curious to trace the progress of plough-making in England. Those of the early cultivators were of necessity rude and imperfect, for, as has been remarked, in those days the ploughman made his own plough. A law of the early Britons in fact directed that no one should guide a plough until he was able to make one. The driver was, by the same law, to make the traces by which it was drawn, and these were to be formed of withes of twisted willow, a long exploded custom; many of the olden terms of which, however, are still retained by the rustic ploughman. Thus the *womb-withy* is yet called the *wambtye* or *wantye*. *Withen* trees are denominated *witten trees*, or *whipple trees*, &c.

It is uncertain whether the early British plough had wheels; some of those of the Saxons were certainly furnished with them. The annexed engraving is taken from a Saxon Calendar. (*Cotton. MS. Tib.* b. 5.)

Yet it is pretty certain that they used ploughs of a form rivalling in simplicity those of modern India; a rude sketch of one of these is given in a Saxon MS. (*Harl. MS.* 603.)

From this cut it would seem that our Saxon forefathers were wont to fasten their horses to the plough by the tail; a barbarous custom, which certainly was formerly practised in Ireland to such an extent that the legislature in 1634 found it necessary to interfere, and by the 11 & 12 Car. II. c. 15, (Irish

Parl.) intitled, "An Act against plowing by the Tayle, and pulling the Wool off living Sheep," declared that "in many places of this kingdome there hath been a long time used a barbarous custome of ploughing, harrowing, drawing, and working with horses, mares, geldings, garrans, and colts, by the taile, whereby (besides the cruelty used to the beasts) the breed of horses is much impared in this kingdome. And also divers have and yet do use the like barbarous custome of pulling off the wool yearly from living sheep, instead of clipping or shearing of them." These wretched practices are then declared illegal, and to be punishable with fine and imprisonment.

The Norman plough was also furnished with wheels, and it was usual for the ploughmen to carry a hatchet to break the clods, as is depicted in the ancient picture from whence the annexed sketch is engraved.

It is pretty certain that the ox was at first, and for a lengthened period, the only animal employed to draw the plough. Thus, although the plough and oxen are so frequently mentioned in conjunction in the Bible, the horse is never alluded to for such an occupation: an old British law forbade the use of any animal except the ox for this purpose. The first representation, of which I am aware, of a horse employed in the plough, is that given (A.D. 1066) in the tapestry of Bayeux.

There are evident traces in the early English agricultural authors of the importance which they ascribed to the improved construction of the plough. This implement, however, was long drawn entirely by oxen in Britain.

Fitzherbert, in his *Boke of Husbandrye* (1532), speaks in a manner that shows that even in his day plough horses were not generally employed; he observes, "a husbande may not be without horses and mares, and specially if he goe with a horse plough." Worlidge, in his *Mystery of Husbandry*, describes (A.D. 1677) very clearly the first rude attempt to construct a subsoil plough: he tells us, p. 230, "of an ingenious young man of Kent, who had two ploughs fastened together very firmly, by the which he ploughed two furrows at once, one under another, and so stirred up the land twelve or fourteen inches deep. It only looseneth and lighteneth the land to that depth, but doth not bury the upper crust of the ground so deep as is usually done by digging." When Heresbach wrote (1570), it was not uncommon in some of the warmer parts of Germany and Italy to plough during the night, "that the moisture and fattness of the ground may remain shadowed under the clodde, and that the cattell, through overmuch heate of the sunne, be not diseased or hurt." (31 b.) Jethro Tull, more than a century since, paid considerable attention to the plough; he had even searched into the early history of this implement, and concluded that it was "found out by accident, and that the first tillers (or plowers) of the ground were hogs." (*Husb.* p. 131.) The ploughs which he describes, and of which he gives drawings, were evidently (although rudely and heavily constructed) superior in several respects to all that had preceded them.

It is not necessary to do more than thus slightly advert to the various notices which are to be found in the early histories and pictures of this invaluable implement; for, in fact, for ages the plough was little more than a rude clumsy instrument, which served only to tear up the surface of the land sufficiently deep for the seeds to be buried. It was not brought to any thing like a perfect tool for the purposes required till the close of the seventeenth century.

The Dutch were amongst the first who brought the plough a little into shape, and by some means or other the improved Dutch plough found its way into the northern parts of England and Scotland. Those who have traced the history of the plough agree that one made by Joseph Foljambe, at Rotherham, under the direction of Walter Blythe, author of some works on husbandry, and for which plough a patent was obtained in the year 1730, was the most perfect implement then in use; and to this day it is well known, especially in the North of England, by the name of the Rotherham plough.

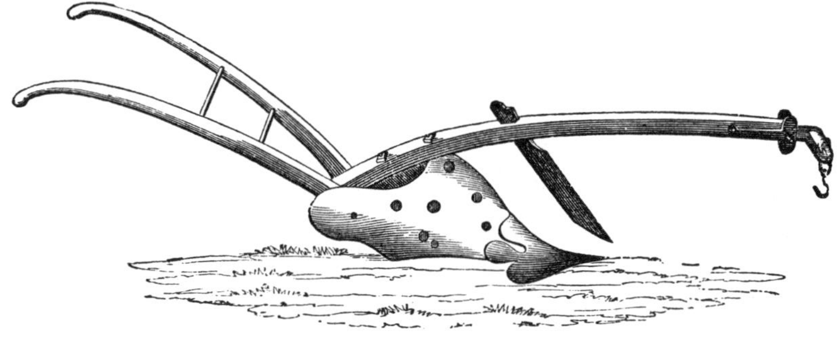

ROTHERHAM PLOUGH.

This plough was constructed chiefly of wood; the draught irons, share, and coulter, with the additional plating of iron to the mould-board and sole, being the only parts made of iron.

Mention must now be made of a step in the march of improvements by the ingenious and justly celebrated James Small, a Scotchman. He constructed a plough on true mechanical principles, and was the first inventor of the cast-iron turn-furrow, commonly called the mould-board; and, although more than a century has since passed, Small's plough may, in most respects, be referred to as a standard for the elements of plough-making.

James Small established his manufactory of ploughs aud other agricultural implements at Black Adder Mount, in Berwickshire, in the year 1763, and died about thirty years afterwards, having devoted the best part of his life to the furtherance of pursuits connected with agriculture.

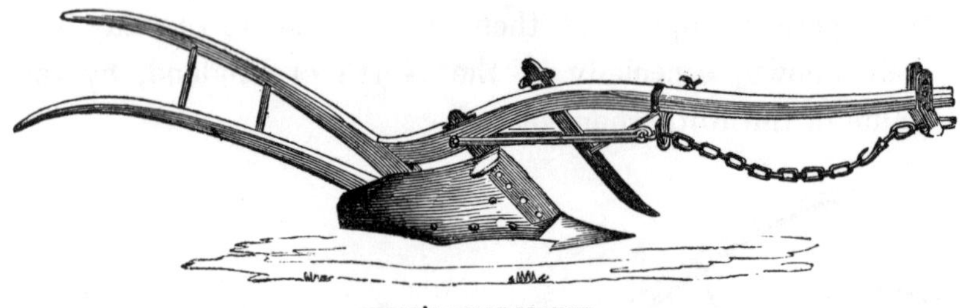

SMALL'S CHAIN PLOUGH.

What is known as the Scotch plough comes next under our view. This comprises the improvements made by Small; but, instead of the beam and handles being of wood, the whole is made of iron, and from the various graceful curves in its outline, it forms one of the most tasteful in appearance, as well as most effective ploughs of the present day.

The Wilkies and Finlaysons, whose names are scarcely less familiar to the English than the Scotch agriculturist, have brought the figure and manufacture of the Scotch swing plough to its present state of perfection; and in *Finlayson's Ploughman's Guide*, and in Stephens's modern work, entitled *The Book of the Farm*, the most perfect form of the Scotch plough is given; and in the latter interesting work its respective dimensions and construction are so clearly laid down, that it may be made therefrom by any skilful workman, unassisted by a sight of the plough itself. The following cut represents this plough.

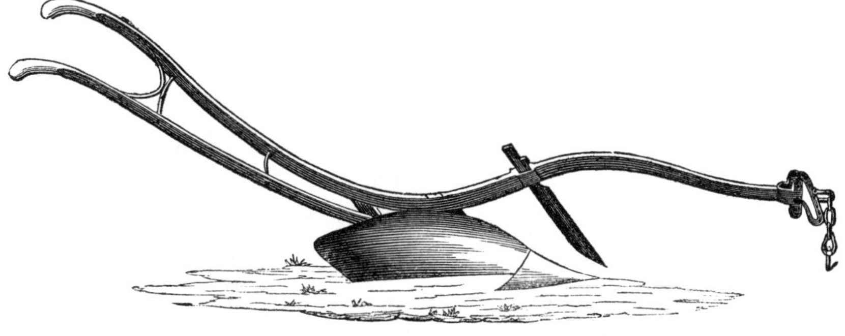

SCOTCH PLOUGH.

Whilst these improvements were going on in Scotland, it will be interesting to observe how similar improvements were progressing in various parts of England.

The Rotherham plough, from its celebrity, had become partially known at a distance from the place of its origin, and was used as a plough better suited than any other for general purposes. The construction, however, admitted of many easy deviations, and, according to the judgment or caprice of various plough-makers, was altered to suit local convenience or prejudice. In most of the English counties ploughs were made similar to the Rotherham plough, chiefly of wood, having the share, coulter, a partial plating of the mould board, and sole, of iron. The plough was generally the joint manufacture of the village wheelwright and blacksmith; and while thus depending upon each other for the completion of the implement, neither of them having the entire control of its form, each made a separate charge for his materials and workmanship, and such continues to be the practice in many districts at the present time. It is easy to imagine that the wheelwright would allow of but little interference in the course of improvement from the blacksmith, and that the latter mechanic would not have his conceits thwarted by the artificer in wood; but rather that each would uphold that state of things which least interfered with the occupations of the other, and best contributed to their individual interest. On the other hand, where either the wheelwright or blacksmith had the entire work in his own hands, a different result was apparent in the good and effective ploughs which were produced by them. In some counties, even to this day, there are wheelwrights who make their ploughs chiefly of wood, and yet have gained, by their peculiar workmanship, celebrity as ploughmakers. The same may be said of blacksmiths; and, as an instance of the skill and merit of one of the latter class, we refer to the account given by Arthur Young, in his *Agricultural Report of Suffolk*, where he mentions a plough made of iron by a "very ingenious blacksmith of the name of Brand," and of which he states there is "no other in the

BRAND'S PLOUGH.

kingdom equal to it." The Report was published in 1804; and at that time, the writer adds, Brand had been dead some years.

The wood plough before mentioned had many objectionable points and properties, as generally constructed. It was very liable to get out of order from exposure to the soil and weather, and continually required repairs. For want of a correct principle or guide, it would frequently happen that if two ploughs were made by the same maker, they would not work exactly alike; and it was a matter more of chance than certainty whether either would perform its work properly.

Ploughshares had been hitherto made of wrought iron, until, in 1785, the late Robert Ransome, of Ipswich, first obtained a patent for making "shares of cast iron;" and this circumstance is worthy of notice, not only as a very important and successful improvement in the part in question, but as the means of drawing the attention of that individual and many others to further improvements in the plough, which were very soon after carried into effect.* In 1803 Robert Ransome obtained a second patent for a mode of applying a case-hardening process to cast-iron shares.

* Of those persons whose ingenuity, spirit, and perseverance, have enabled them, through many opposing difficulties, to effect improvements in Implements of Agriculture, I desire to make, though briefly, the honourable mention they deserve; of this class I have already named, Foljambe, Small, Wilkie, and Finlayson; and, although I may be thought to have the partial feeling which attaches me to one I loved and honoured in his day; yet I cannot on that account forbear also to mention the late Robert Ransome, my grandfather, who commenced about sixty years ago a series of experiments in reference to Implements of Agriculture, which he carried on with determined perseverance till he accomplished the object he had in view. The manufactory at Ipswich, now carried on by J. R. & A. Ransome, was established by him.

This invention is now so well known that a brief description of it may suffice.

Before the time referred to, cast-iron shares, although occasionally used in some districts, were found to wear away too fast from the under side. When the first edge was worn off, the share became too thick to cut the ground properly, and its tendency, when so worn, was to "lose its hold of the work," and to pass over weeds without cutting them; while the blunt edge greatly added to the draught. The improvement alluded to is that of case-hardening the under-side the thickness of one-sixteenth or one-eighth of an inch, which is, in effect, like a layer of steel underneath the share. This part, from its hardness, wears very slowly, while the upper part of the share grinds more quickly away, thereby producing a constant sharp edge on the under side. The land-side point of the share is also hardened in a similar manner, which prevents it from wearing so fast as it would otherwise do at that part; a tendency to which all shares are more or less subject. The following figure shows a broken share, in which the white lines indicate the hardened parts.

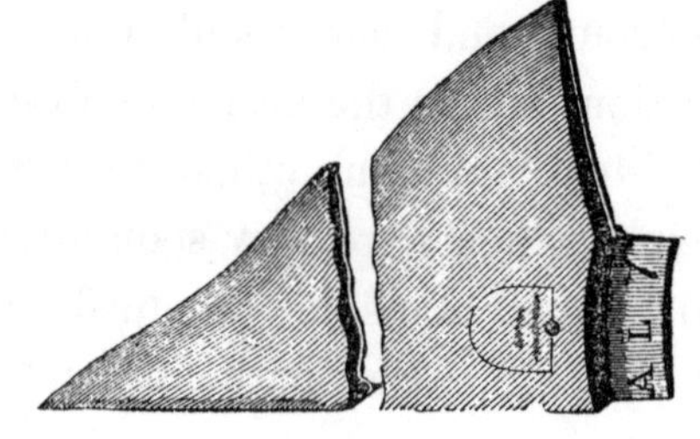

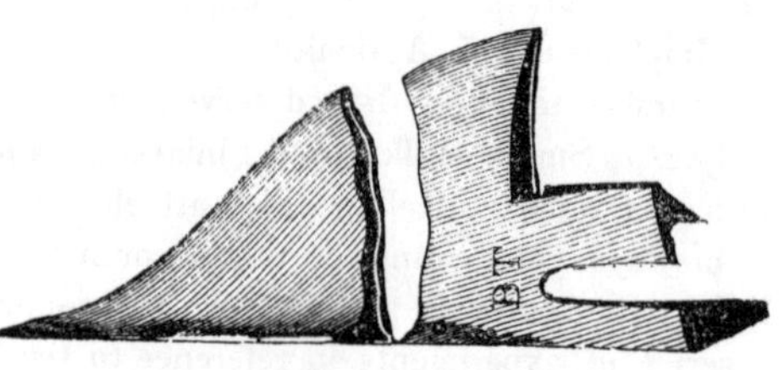

By the use of cast-iron shares, thus tempered, considerable expense is saved; the first cost being so much less than those of wrought iron, that they may be renewed at smaller charge than the repairs needful to the latter. Another benefit arises from their use, that the consumer is no longer liable to those inconveniences to which he was before necessarily subject with wrought iron shares, from the frequent repairs requisite in sharpening and "laying them;" an operation requiring the aid of a blacksmith, who, generally working at a distance from the farm, very often unprofitably occupied the time of the men and horses.

Following up this improvement in the shares, a Suffolk farmer invented for his own use a cast-iron plough-ground or bottom, with a moveable sole or slade; this had mortices to receive the tenons of the wood, to which it was attached. This plough-ground soon became of general use in the counties of Norfolk, Suffolk, and Essex, and but few ploughs were then made without it.

PLOUGH-GROUND AND SHARE.

There was still a defect to which even ploughs made with this iron-ground were liable, which was, that nearly the same uncertainty attended their manufacture as in those constructed entirely of wood; scarcely two workmen would make them alike, and sometimes one plough would work well and easy to the holder, while another made by the same hand would be inferior in these respects; in addition to which inconvenience, the wood tenons in the iron mortices were liable to decay, and a constant expense of repair was entailed, which has of late years, under further improvements in the construction of ploughs, been obviated,

As improvements in the art of founding became known, cast iron gradually, to a great extent, superseded, as a material for the manufacture of ploughs, the use of wood and wrought iron, from the facility and economy with which parts requiring nicety in their form could be multiplied to any extent, with the certainty of their being always alike. Soon after the introduction of the cast-iron share, ploughs were invented having their entire bodies made of cast-iron.

The frames of which the three following are sketches,, were so contrived as to admit of the handles, beams, shares mould-boards, soles, and other parts being screwed to them, and any portion altered or removed at pleasure. They also admitted of the mould-board being set to a wider or narrower furrow, and of changing the shapes of the different parts as they were required for different purposes.

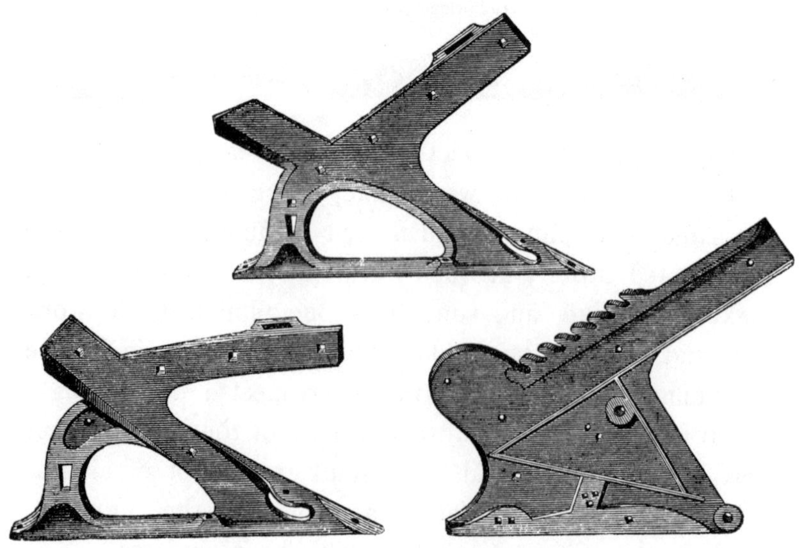

PLOUGH.FRAMES.

The following figure shows the entire body of the plough.

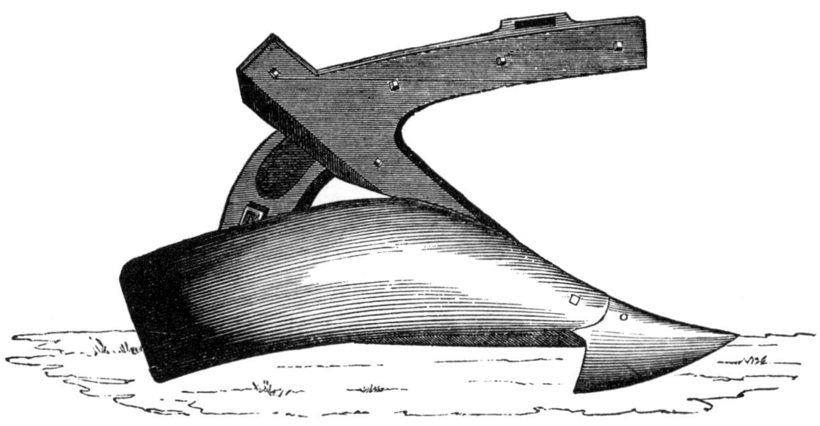

PLOUGH BODY.

As every part of the plough upon this construction admits of being easily replaced by the ploughman without the aid of a mechanic, the farmer has only to keep by him a stock of the wearing parts, to insure his plough being at all times in working order, and in the original form.

This arrangement of the body of the plough was a very considerable improvement beyond any thing that had been before produced, and is applicable to ploughs of almost every description, as has been fully evidenced in the exhibitions of the different plough-makers, at the annual meetings of the Royal Agricultural Society, almost every one of whose ploughs being so constructed.

Having thus far traced the several improvements which have been made in the construction of the different parts of the plough, and having given a reference to and figure of a Scotch swing plough (see p. 15.), I shall here introduce a cut of one of the most approved English swing ploughs, and remark on the mechanical principles on which the swing plough should be constructed; and, in pursuance of this intention, shall commence with

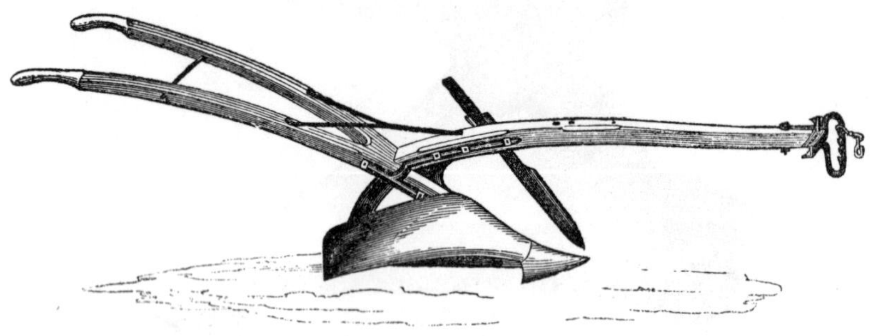

ENGLISH SWING PLOUGH.

The Plough Handles.—The handles should be sufficiently wide apart to allow the ploughman to walk in the furrow, and long enough to give him a full command of the plough, so that he may be able with ease to lift or depress it readily in work, bias it to the right or left hand, and swing it round at the land's end from one furrow into another.

The Plough Beam.—This should be of such a length, that its end, commonly called its head, should cut at the point of draught, upon a line drawn from that part of the collar to which the traces are attached, to the share or that part of it where it first raises the soil, as may be seen by reference to the engraving of the wheel plough with high gallows, at p. 34. On the right arrangement of the point of draught in the structure of the plough depends much of its steady working at its proper depth. It is from the principle of balancing from a point adjusted to the line of draught, that the plough takes its name of *swing* in contra-distinction to the names of *foot* and *wheel* ploughs, which will be hereafter described.

The beam should be curved upwards at the coulter and throat of the plough, to clear itself of rubbish which sometimes accumulates, and should be inclined slightly from the land, or, in other words, towards the furrow, because its tendency is to yield toward the loosened land, and it therefore requires this counteraction in the line of draught to keep it in a right direction. This is supposing a pair of horses to be harnessed abreast; if they are harnessed at length, the beam should be still more inclined; for, as neither horse then walks on the "land," the direction of the force towards the land-side is still further decreased.

The Plough-head.—The cross head of the plough, as shown below, forms a ready means of increasing or decreasing the inclination last spoken of, and the hake, or draught iron, which moves in the arc of a circle along the cross head, has notches by which the depth of the plough can be regulated in unison with the line of draught. There are various contrivances for these purposes, most of which involve the use of a screw as a means of adjustment; but the plan of pins and notches is sufficiently accurate, and not liable to be out of order. The following are sketches of two, the one English, the other Scotch, which appear unobjectionable.

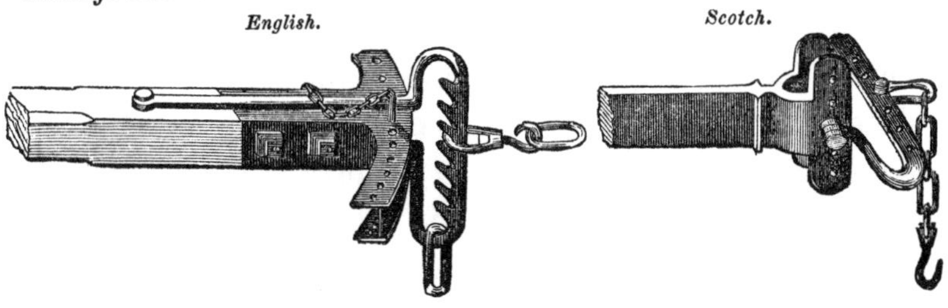

PLOUGH-HEADS.

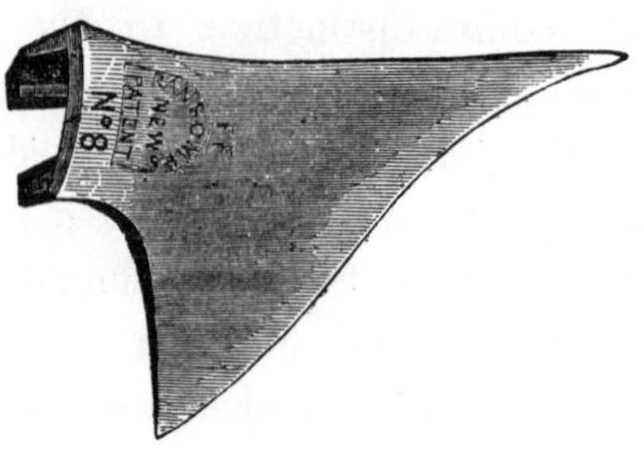

The Plough-share.—The plough-share is that part which cuts the soil horizontally, and is variously shaped for different purposes. On stony lands it is best with a point; but where the land is free from stones, the wing is best when angular, and the cutting edge in a line, or nearly so.

For hard lands the share requires to have a greater "pitch" or downward inclination towards the point, and it is common for lands in this state to use new shares and to wear them a day or two, and then lay them aside for "summer lands."

The Mould-board.—The upper part over the box of the share should form the first part of the rise of the mould-board. After the coulter and share have made the vertical and horizontal cuts for the depth and width of the furrow-slice, the mould-board has to complete the work by turning it over and leaving it in its proper position. On the precision with which this part of the plough performs its work, much, indeed nearly all, of the beauty of the ploughing depends: hence the importance of discovering its true form for the land on which it has to be used. Desirable, however, as this is, there does not as yet appear to be any precise rule for the formation of the mould-board, that has met with so uniform an approval under the test of practice, as would lead me to speak with entire confidence of it. I have looked at the mechanical principles laid down by Small, Bailey, Gray, Amos, Jefferson, Clymer, and others, but am not aware of any plough-makers of the present day who strictly adhere to either the one or the other; and so long as the mould-board cannot be used on even the same farm under circumstances always similar, as its operation will necessarily be affected by the weather, the state of the land, with the varying depth and width of the plough, it is not an easy matter to determine which form is best for general purposes. It is clear that different soils, as, for instance, light sand, and heavy clay, require mould-boards almost the opposite of each other. The Norfolk mould-board, for instance, is short, with rather a hollow or concave surface, whilst that used in the hundreds of Essex is long and convex.

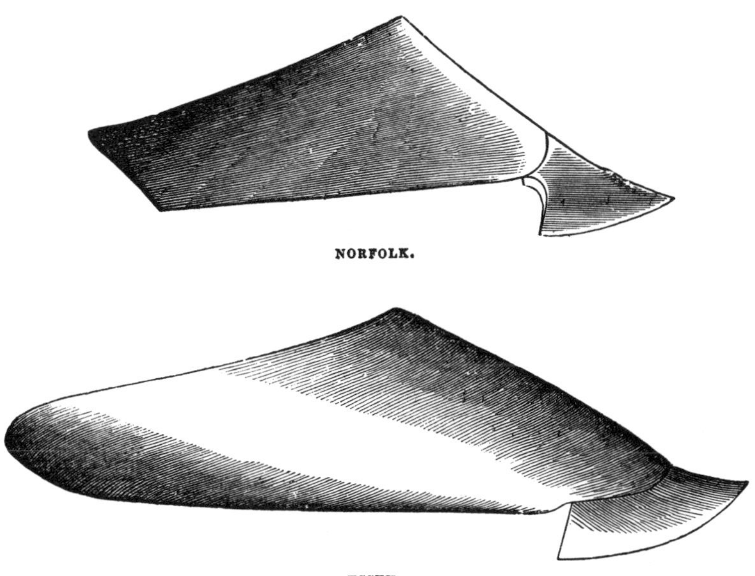

NORFOLK.

ESSEX.

Lord Western, many years ago, improved upon the form of the heavy-land mould-board in use in that part of the county of Essex where he resides, by cutting away a considerable portion of the lower and hinder part of the figure, and by making it in a straight line lengthwise from the nose to the hind part.

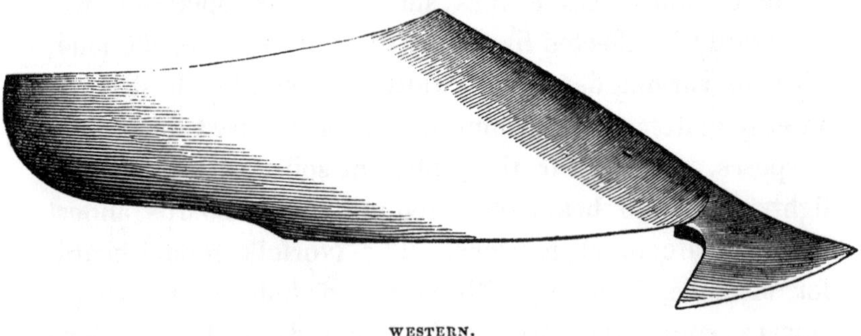

WESTERN.

Were the circumstances always the same, there can be no question but that one mathematical form of the mould-board would be preferable to all others; but, under circumstances so various, the plan hitherto adopted has been to prove, by experience and from practical operation, the forms best suited to different lands under an average depth and width of work, keeping as nearly as possible to the principle of the wedge, as necessary for the proper lifting, turning, and laying over the soil. Provided the mould-board be made so that the work, while in operation, is properly performed, a good practical criterion as to its figure will be found in the evidence of friction it has undergone, and this, with the fine cast metal now in use, can be determined to a nicety. If, on a given soil, the mould-board becomes uniformly brightened—if the mould appears to slip with light friction and with the same pressure from one end of it to the other, it cannot be far, if any thing, out of its proper shape for the purpose intended. And yet, if the same mould-board be used on some other lands, it will immediately show its inapplicability to them by the soil adhering to it in parts, not slipping well through it, and thus evidencing a want of uniformity in its general friction. Therefore, considering that neither depth nor width of furrow is always the same, and that scarcely one circumstance affecting its use is unvarying, it is difficult to find a rule which shall aptly suit these changes. At the same time it is not presumed that such will not or cannot be found; and the theory which most accords with my view, is one which has recently been laid down by W. L. Rham, rector of Winkfield, Berkshire, a gentleman whose scientific and agricultural knowledge entitles his opinion to considerable deference. His theory is, that the mould-board should be composed of straight lines in the direction of its length, with continually increasing angles to the line of the furrow; these last lines being either straight, convex, or concave, horizontal sections of the mould-board would then exhibit lines of this kind—

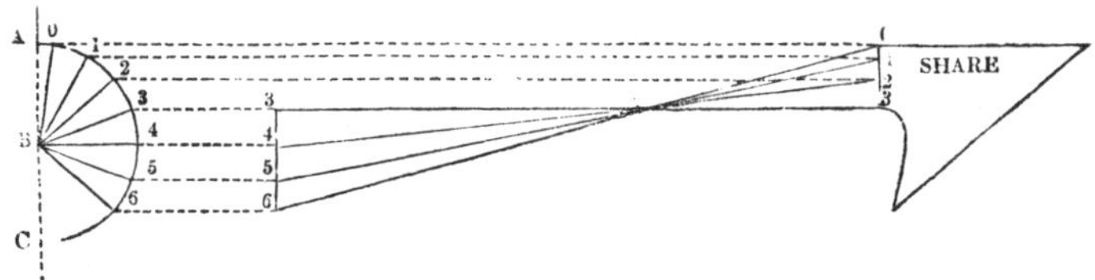

whilst the perpendicular sections would be thus—

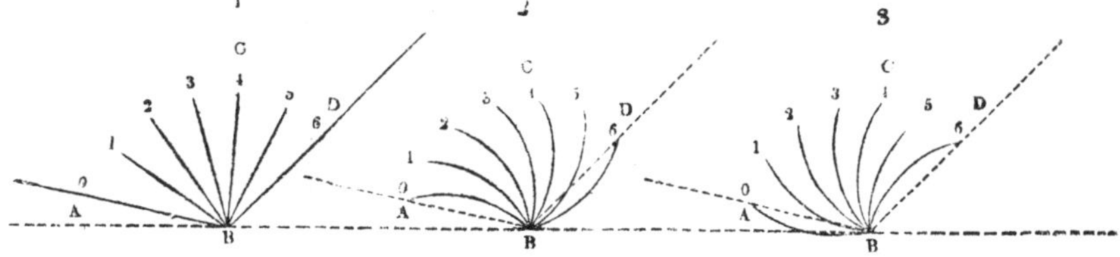

A B being a section of the mould-board just behind the share, C B in the middle of its length, and D B at the heel. *Fig.* 1 is a section in which the vertical lines are straight, and is adapted to mellow soils; *fig.* 2, convex lines for adhesive soils, and *fig.* 3, for very light loose sands. This subject is more fully explained under the head "Plough" in the *Penny Cyclopædia;* an article which I would recommend to the perusal of my readers.

The following drawing exhibits the operation of the mould-board in the process of turning the furrow-slice.

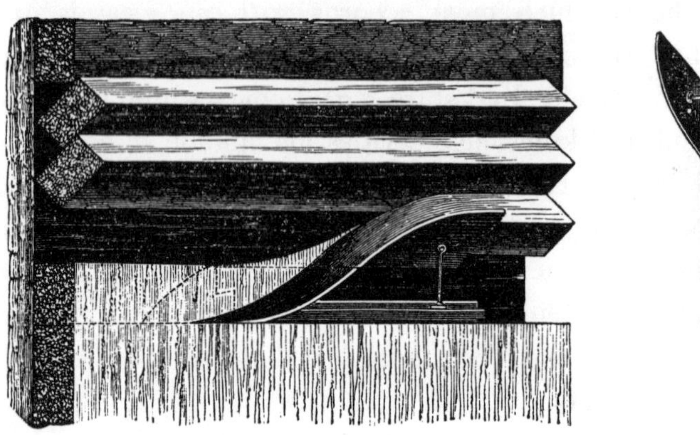

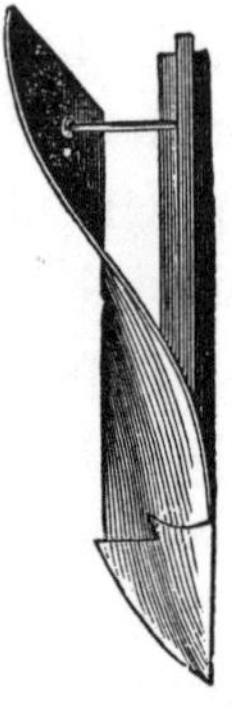

Having availed myself of the opportunity of testing several forms of mould-boards which have obtained general approval in various localities, I find that, though differing widely from each other, their form is nearly that which this rule, under the modifications of its straight, concave, and convex lines, leads to; and the conclusion to be thence drawn is, that although no one form of mould-board will or can be applicable to every variety of soil and circumstance, there is no description of soil for which a perfect mould-board may not be made by this rule in some of its modifications.

There is another description of mould-board, which, from being divided into parts lengthwise, approximates in some degree to the upper and lower rests of the Kent turn-rest plough. The upper part of this mould-board is nearly flat, and stands out in a wedge-like form to turn the furrow; the lower part is called the ground-rest, which is fixed below and within the upper part, and cleans out the furrow.

This kind of mould-board was originally made by Pritchett and Perry, of Millbrook, near Southampton, in the year 1809, and has continued in use ever since. It is the original of one described by Philip Pusey, Esq., in an interesting paper reporting his "Experimental Enquiry on Draught in Ploughing," as used on Hart's plough in a somewhat improved form. (*Journal of the Royal Agricultural Society*, vol. i. p. 222.)

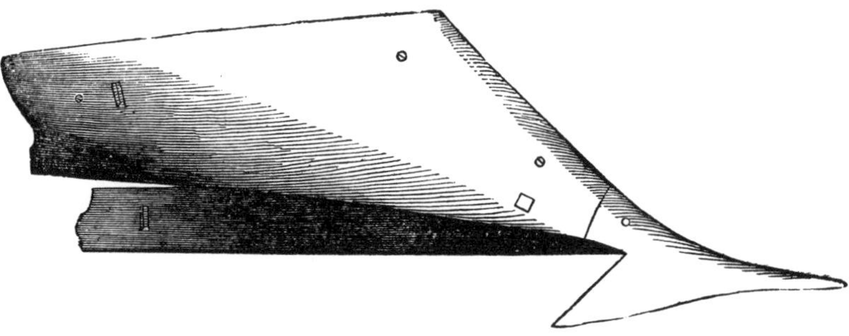

The Coulter.—Simple as the coulter may appear to be, it is a very important part of the plough, and much depends upon its being properly formed and fixed for the work it has to perform in the operation of ploughing. It

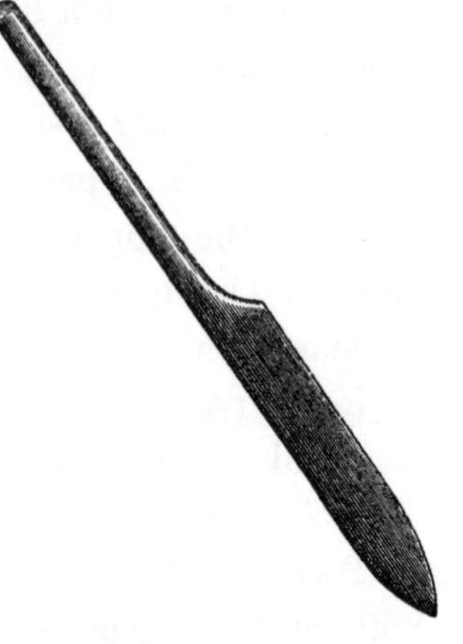

should be made of iron and steel, and of sufficient substance to stand firmly to the position in which it is set for its work, not to bend either to the right hand or to the left. The blade or cutting part should be about two inches and a half wide, and formed by the meeting of two curves shown in the accompanying figure, as this shape cuts the land easier than when the edge is either in a straight line or curved forward. The land side of the coulter should be flat, and the opposite side a gradual taper from the edge to the back : the thickness must be determined by the strength of the work it has to perform.

The angle at which the coulter is usually set, is about fifty-five degrees with the plane of the ground ; but in "summer-lands" it requires to be placed in a more slanting position, and to take the lead of the share about three quarters of an inch, to prevent the grass or rubbish driving in a heap, as it otherwise might do. On the contrary, when used for ploughing up hard fallows, it requires to be fixed in a more upright position, and rather more backward than the point of the share. It should be placed about half an inch above the share, and a quarter of an inch to the land side of it. Every good ploughman has his own notions on the subject of setting the coulter, but the above directions are given from practical observation.

The usual mode of fixing the coulter in the socket of the frame or beam, is by means of wood or iron wedges driven above or below the socket, or by a coarse cut screw-bolt, which turns into the side of the socket and presses against the coulter stalk. Each of these modes is defective. The plan of wedging the coulter is difficult, and the ploughman, after trying his skill for some time, frequently finds himself foiled in his attempt to set it in its proper position. The plan of the screw-bolt just described is also faulty, for wedges are even then requisite, and the difficulty of setting is very little, if any, less than by the other mode.

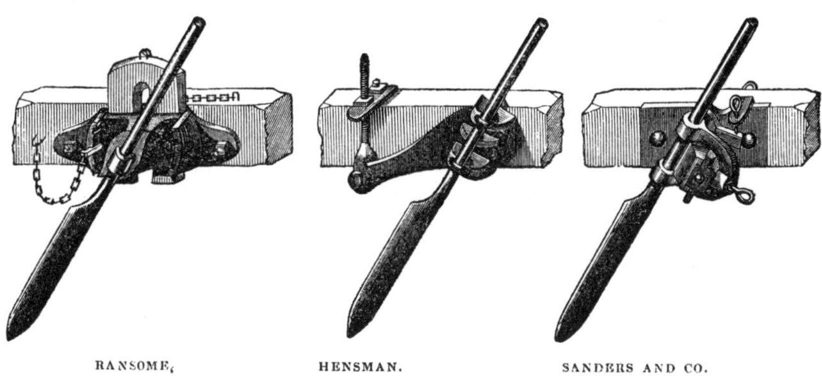

RANSOME. HENSMAN. SANDERS AND CO.

The preceding cuts exhibit some of the contrivances to obviate the difficulties referred to. Although none of these is quite free from objection, they, however, admit of the coulter being easily raised higher or lower; the angle of inclination as readily altered; the edge of the long blade turned more towards or away from the land, and the coulter set out or in laterally.

The first was made at the suggestion of George Green, a spirited farmer at Millbrook, in Bedfordshire, and was followed by Sanders and Co., of Bedford, and Hensman of

Woburn, with their patent modifications of the same principle.

Skim Coulter.—This is in principle a miniature plough in the usual position of the coulter, for the purpose of cutting the surface grass and rubbish, and laying it in a position to be completely buried by the furrow slice, which is turned over by the plough. This addition is not in very general use, but it produces very neat work.

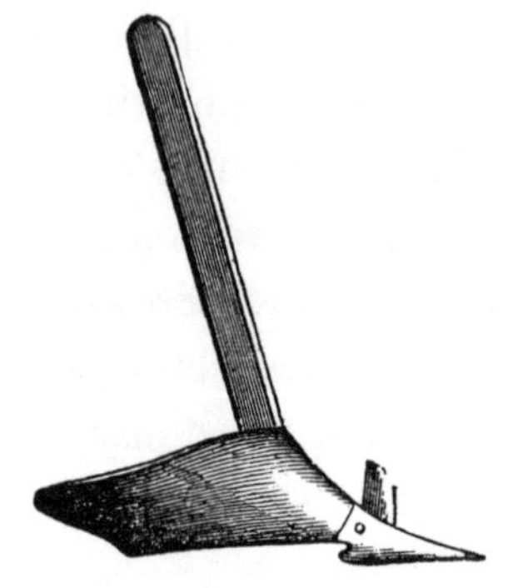

SKIM COULTER.

Wheel Coulters are used in the fenlands, and are useful when ploughing up turf. One of these coulters is represented by the annexed figure. The cutting disc should be made of steel, with a nave sufficiently long for it to be steady, and the box should be bored true, and revolve on a well fitted steel pin, as on the perfect fitting of the box and axle the correct working of the disc, and consequently the effective operation of the plough, depends.

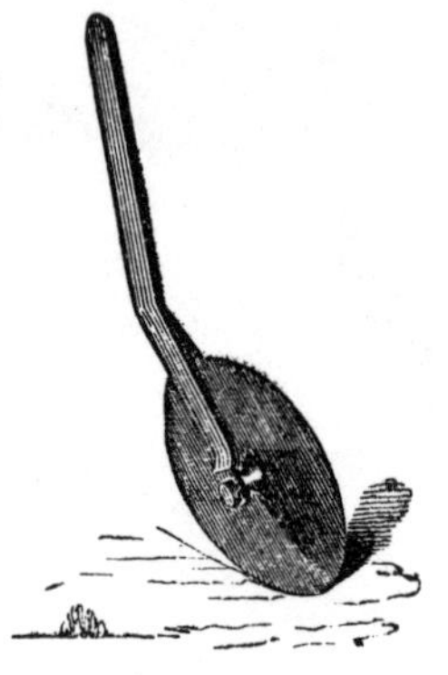

WHEEL COULTER.

Having thus far described the separate parts of the swing plough, it now only remains for me to remark upon the comparative advantages attributed to its use.

1. *It admits of being set into its work at a given depth, either shallower or deeper, by the mere alteration of the draught iron at the point of draught, or by increasing or*

decreasing the distance at which the power of the horses is applied.

2. *The ploughman has also the power of regulating, in some degree, the depth of the work, by either lifting or bearing upon the handles.*

3. *Its construction is more simple than that of the wheel plough, and less expensive in its first cost.*

4. *A skilful workman can plough across the ridge and furrow at very nearly an uniform depth; he can work with it on almost all descriptions of land, and in all weathers when ploughing can be accomplished.*

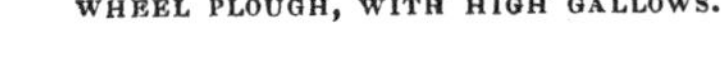
WHEEL PLOUGH, WITH HIGH GALLOWS.

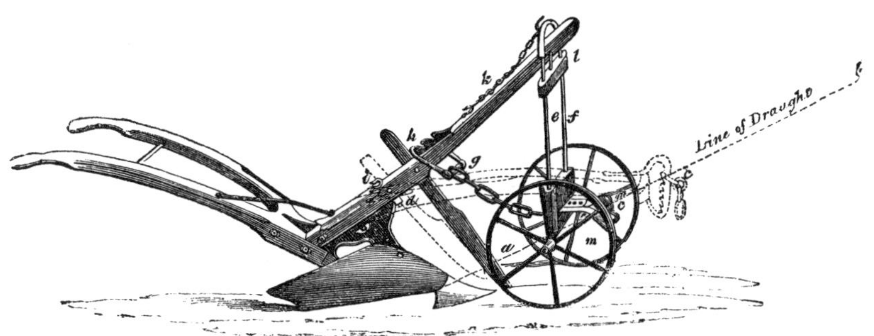

The Wheel Plough (*with high Gallows*).—This implement derives its name from having the appendage of a carriage and wheels. The body of this plough is essentially the same as that of the swing plough, and notwithstanding the different form of its beam, the point of draught should be the same as that of the swing plough, namely, to cut a line drawn from the horse's shoulder to the share or point of resistance. Its form will be seen from the following sketch of the plough used in many parts of Norfolk, Suffolk, and

Essex; and as all ploughs of this class are the same in mechanical principle, the one we have chosen for illustration will suffice.

The dotted lines upon the figure, from the frame at *d* to the point *c*, will give the figure of the swing plough; and what is intended by this, is to show that the line of draught will be the same in either case, and that the line *a b* should be intersected by the draught-iron of the plough.

The beam of the wheel plough is elevated and made to rest on a bar of wood called the bolster-crossing, the upright standards *e* and *f*, and the latter form part of the carriage framing supported by the wheels; the draught-chain *g* collars the beam at *h*, and will remove at pleasure from the notches at *h* to the one at *i*; the small chain *k* serves to keep the standards in their upright position; *l*, the bolster, is made to rise and fall as the plough may be required to cut shallower or deeper. The plough is made to go more "to or from land," that is, to the right or left, by altering the chain to a notch in the hake at *g* either to the left or the right hand, or the same at the hake *m*. The former acts instantly on the body of the plough, the latter on the carriage and wheels. Corresponding alterations with the latter must be made by the pins in the bolster.

It would appear that the point of draught in this plough should be the same as in that of the swing plough, but a slight deviation from this is adopted in practice, from the unevenness of land rendering it desirable to bring a slight pressure upon the wheels, in order to insure an uniform depth of furrow, and this is effected by making the point of draught slightly above the part it would cut on the direct line; it will then be obvious that any remarks which apply to the one plough, such as pitching the share more or less into the ground, varying the breadth of the furrow by expanding or contracting the width at which the mould-board works, or similar alterations, will apply equally to the other. The difference, therefore, in the construction of the wheel plough and the swing plough lies in the addition of the carriage and wheels, and those parts which are in direct connection with them.

By a reference to the cut of the wheel plough, and description of its several parts, it will be seen how each part of the wheel gear mechanically operates; but, it may be observed, that irrespective of any alteration in the dip of the share, as that would alike apply to either the swing or the wheel plough, the mode of setting the wheel plough at a greater or less depth, may either be accomplished by lowering or raising the cross-bar *l*, or by increasing or decreasing the length of the top-chain *k*; but as this chain is chiefly intended to keep the gallows in an upright position, this mode of altering the depth of the plough is only resorted to when alterations are suddenly required.

The draught-chain, which is attached to a notched loop or hake, passing round the beam of the plough, admits of change to different points of draught, for the purpose of fixing it at the point at which the plough balances itself best: if it be set too forward, the plough will incline too much towards the ground at the share, and rise up at the heel; and if too far backward, the converse of this will be the case; the ploughman, therefore, ascertains by trials the point of draught, and regulates this chain accordingly. The advocates for the wheel plough state—

1. *That ploughing can be more easily performed by this than by the swing plough.*

2. *That the work can be executed with greater precision, both as to depth and width, inasmuch as the wheels serve as gauges.*

3. *That shallower ploughing can be effected with the wheel than with the swing plough, and that when the ground is very hard or very stony, it can be worked with more ease, and under some circumstances, in which the swing plough, cannot be held in its work at all.*

The advocates of the swing or the wheel plough, each advance good reasons for the use at particular times, or on particular soils, of the one or the other; but my object is to furnish a clear description of both, and to point out the mechanical principles, by which their respective merits may be compared.

The swing plough requires more *skill* than *judgment* in the user. It is dependent upon the ploughman for almost every inch of its work; it will not go far without his aid, and long practice only can make him perfect in the use of it; but, when he becomes master of the art, he can cross ridge and furrow, and gauge his work with astonishing correctness.

The wheel plough with the high gallows is complicated in its construction, and requires more *judgment* than *skill* in the ploughman, because the depth of its working is not alone dependent on the principle of balancing, but on the adjustment of the wheels and draught-chains, which should act in unison. It should, however, be observed, there is great liability that the forces, which act so well together *while in unison*, may be through the ignorance or carelessness of the ploughman, set in some degree *opposing to each other*, and then the resistance of draught is greatly increased. Whatever approximates to this disagreement of forces is in its meaure an evil; and it is the more likely to be fallen into, as the plough will hold steadily to its work, notwithstanding this mismanagement. Its proper working is not so dependent on the manual dexterity of the ploughman as the swing plough is, because the wheels serve as a gauge for the depth, and the one which runs in the furrow acts as a gauge for the width of the furrow-slice. Some mechanical knowledge must be added to practice, for its proper management; and to cross high ridges and furrows to advantage, the wheels must be frequently altered.

In ploughing across the ridge with a swing plough, a skilful ploughman is able to dip into the furrow, and to rise again on the ridge, by his own command of the plough by the handles; but the wheel plough, unless altered when arriving at the above inequalities, leaves the lowest part unploughed, because the wheels, as they rise at the opposite ridge, must in some cases take the plough nearly out of the ground.

There is generally a perceptible difference in the appearance of the ploughing done by the swing and by the wheel plough, the former not being so evenly gauged and uniformly laid as the latter. When the draught-irons and wheels are all set properly in unison with each other, so that the plough has no conflicting forces in operation, and the wheels press but lightly on the ground, the force required to draw the wheel plough is less than that of the swing plough, from the regularity of its progress throughout as regards both width and depth of furrow.

As a counterbalance to the advantages of the wheel plough, in the precision of the work done, in the saving of labour to the horses, and in the adaptation to hard land;

is the time consumed in frequent adjustments of the points of draught,—the probabilities of these points as forces being set in a counter direction to each other,—the greater complication in its construction and its greater cost, with the liability of the wheels to clog in wet weather; and its inapplicability to uneven land. Balancing all the circumstances, *pro* and *con*, it appears to me that, in comparison with the wheel plough with gallows as described, the practical advantages upon the whole are in favour of the swing plough.

The *Plough with Land and Furrow Wheel* comes next in course, which is another description of wheel plough, and may be considered as a combination of the two before described. This possesses advantages over either of the others, inasmuch as the same plough can be used occasionally as circumstances may require it, either as a *swing* or as a *wheel* plough, and being simple in its construction, easy of management, and is adapted to the ready instruction of boys in the art of ploughing. I have not been able to ascertain, with sufficient correctness to speak with certainty, the county in which this mode of using the wheels originated, but they were attached to the double furrow plough invented by Lord Somerville, and the plan has obtained extensive use, particularly in the midland counties.

I select for illustration a plough with a body of the same mechanical construction as those of the swing and wheel ploughs last referred to, the wheels being attached to the beam in a very different manner from the wheel plough with high gallows. Of late years many Scotch ploughs have been made with wheels fixed in the manner about to be described, and the plough so altered goes by the name of the *Improved Scotch Plough;* but strong and constant as has ever been the attachment of the Scotch ploughman to the swing plough, I have never found a Scotch advocate for the use of wheels. The following cut represents a plough, which, from its having been noticed at several public trials both in England and in Scotland, is the more eligible for the purpose of description.

THE RUTLAND PLOUGH.

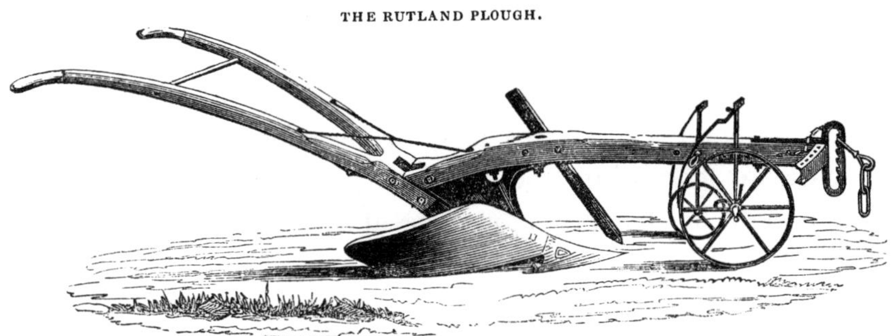

It will be observed that the two wheels fixed at the fore part of the beam, constitute the difference between this and the swing plough. One of the wheels, about twelve inches in diameter, is fixed on the land side of the plough, and runs upon the unploughed land; the other wheel, about twenty inches in diameter, is on the opposite side, and runs in the furrow. The latter wheel is upon a sliding axle, which admits of its being set to any width of furrow. The upright shanks regulate the depth by means of screws and sockets on the beam.

* This plough takes its name from the county, to which it was introduced by Richard Westbrook Baker, Esq., of Cottesmore, for whom it was in the first instance expressly made on a simple plan, suggested to the makers by that gentleman, and is now in general use.

All that has been previously said in favour of the wheel plough with high gallows, may be said of this; but it is more simple in its construction, and, if requisite, the wheels may be taken off, and the plough used as a swing plough without them.

It has been objected that the wheels require frequent adjustment, which occasions loss of time, and that unless the furrows be ploughed beyond the length required, the large wheel must be raised at each end of the field just before the plough comes out of the furrow, or it will be taken gradually out of the ground, and the land will not be ploughed to its full depth. The usual plan is to extend the common furrow two feet, or thereabouts, beyond its ultimate length into the headlands, and afterwards to set this right by the cross ploughing at the top and bottom of the field.

The loss of time involved by alteration of the furrow wheel may be overcome by a simple mechanical contrivance; with a lever, the longer end of which reaches the handle of the plough, and by it the wheel can be adjusted to any depth instantly. The invention of the late Henry Osborne, a Suffolk farmer, effects this purpose, and answers admirably.

LEVER PLOUGH.

In the *Report to the Board of Agriculture* from the county of Leicester, published in 1808, it is stated, " that more than thirty years ago, wheels were first applied to the fore end of the beam, and it was found, by ' pitching ' the plough a little deeper, and setting the wheels so as to prevent its drawing too deep, the wheels were a sufficient guide, and the plough required no one to hold it except in places of difficulty." If a plough with land and furrow wheel be properly adjusted, a lad of fourteen years of age can manage it easily; and I once saw, at a ploughing match, a lad, having the only one of this kind in the field, walking frequently at leisure beside it, to the great astonishment of his many competitors with other ploughs, and from whom, to their still greater astonishment, he carried away the prize. This lad had been taught ploughing only a few months.

When one wheel only is attached to the plough, some persons give the preference to the small one to run upon the unploughed land, as it is less likely to clog up, and requires no alteration towards the end of the furrow; but others prefer the larger wheel which runs in the furrow, as it has an even surface to travel over, and at the same time correctly regulates the width of the furrow-slice. It also more effectually facilitates the turning round at the head land, particularly if the horses have to go to the right hand. The larger wheel to run in the furrow, therefore, is best for general purposes, and with a lever attached to it, as described, it is rendered very easy of adjustment.

In the use of a gauge for the depth of ploughing, whether of two wheels, one wheel, or a foot*, the plough

* By a foot is intended an iron shank, bent at the lower part in a contrary direction to the movement of the plough, used sometimes as a substitute for the wheel.

should be so regulated as to press but lightly on the ground when passing over it; thus admitting as little counteracting force between the wheel and share as possible.

In the Prize Essay by Henry Handley, Esq., the advantages of wheels are clearly set forth, and his arguments in favour of their use have since been very strikingly confirmed by the trials made under the direction of Philip Pusey, Esq. (*R.A.S.E. Journal*, vol. i., pp. 140, and 219.)

PLOUGHS WITH A FRICTION WHEEL.

The four following sketches of ploughs, each with a friction wheel are not dissimilar in principle, though slightly different in the mode of attaching the wheel.

WILKIE'S, 1814.

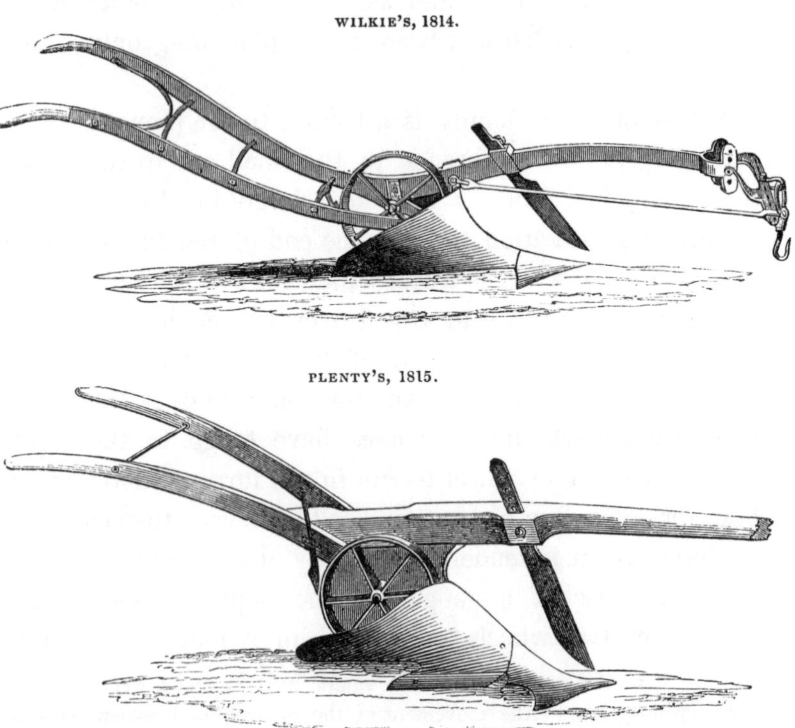

PLENTY'S, 1815.

M'CARTHY'S, 1817.

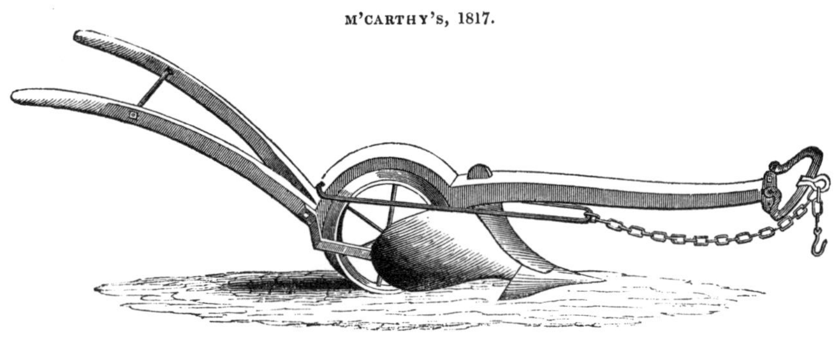

PALMER'S, 1840.

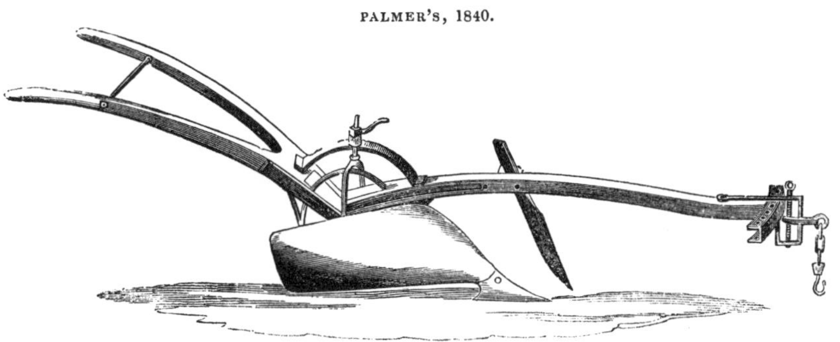

The intention of placing a wheel in the position shown in the hinder part of the plough, is to reduce the friction that arises from the sliding pressure of the sole, and in the case of that of Palmer's, for the further purpose of regulating by a screw, which alters its depth, what may be termed the under "pitch" of the plough, so as to make it take more or less "hold of the ground."

There appears something plausible in the theory of thus reducing the friction at the sole, and of making the plough into a sort of wheel carriage, by which its motion forward may be rendered more easy of draught; but I am convinced these are more than counterbalanced by the disadvantages which attend it in practice, viz., The increased weight—the wheel getting loose in wear, and becoming unsteady at the axle, so as to rock a little on either side, clogging upon some soils,—and the complication it gives to the structure of the plough.

If a plough be "set" properly, there is but little friction upon the sole at the heel, and the saving in that respect bears a very small proportion to the entire draught of the the plough. It will be seen by the following diagram, that the plough, as commonly constructed, bears only upon the edge of the share and the heel, not all the way upon the sole.

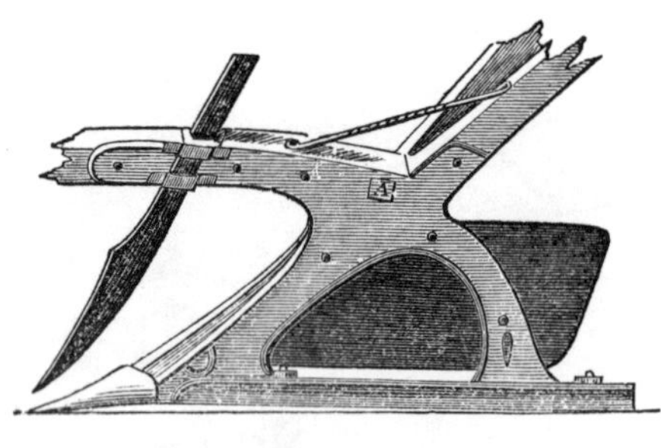

The alteration of the pitch of the plough applies equally to a sole made to rise or fall as to a wheel, and the bearing on the ground may be made as short with the one as the other.

DOUBLE FURROW PLOUGH.

This plough was the invention of Lord Somerville, and

is for the purpose of ploughing two furrows at the same time. There has been no very material alteration in the principle of this plough; but as regards the style in which it is now made, it has progressed in the same degree as the common plough, and is a neat and well-finished implement.

On light soils, where there are no stones to throw either of the ploughs out of work, and where not more than three horses are required, these ploughs may be beneficially used, and a considerable saving effected in the cost of ploughing.

A farmer residing in the neighbourhood of Ipswich, has three double furrow ploughs at work, and the account he gives of them is—

"That with three horses abreast, and one man only, he can plough double the quantity of land per day that he can with a single pair horse plough; in other words, he ploughs, with his double furrow plough, two acres instead of one: thus he saves a man and a horse. He admits it may be rather more work to the horses, but not worth mentioning. He ploughed thirty acres in fifteen successive working days and the horses did the work without being in the least hurt by it. The ploughmen did not at first like these ploughs, but they soon became reconciled to them. The land in which they were used is light, but inclining to what is termed a mixed soil or sandy loam. The furrows are nine inches wide by six deep."

It will be recollected that the silver medal of the Royal Agricultural Society of England was awarded to John Clarke, of Long Sutton, Lincolnshire, for the invention of this plough. (*Journal of Roy. Agr. Soc.* vol. i. p. 66.)

CLARKE'S UNIVERSAL PLOUGH.

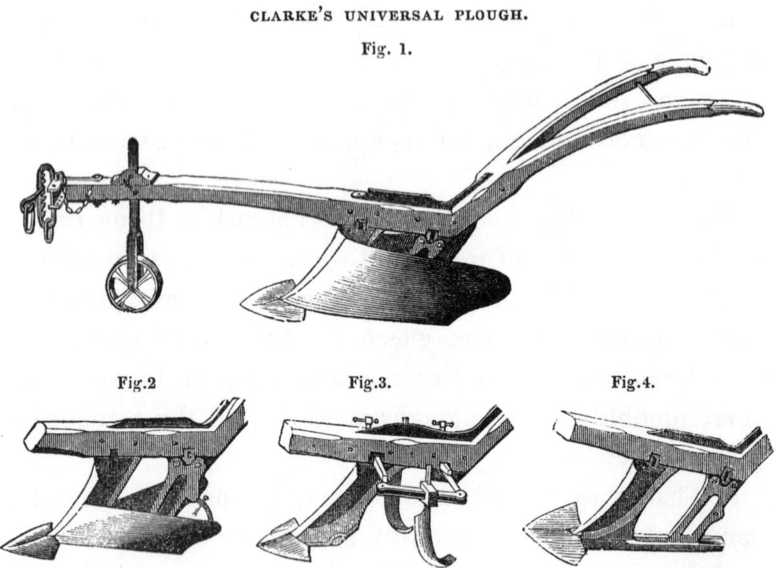

It is for the purpose of ridge culture; and by an easy transition of shape, which is accomplished in a simple manner, it becomes—

1. A double tom or ridge plough.
2. A moulding plough.
3. A horse-hoe, or cleaning plough.
4. A skeleton, or broad-share plough.

The above cuts illustrate the changes of form of which its construction admits.

Although I have already alluded to the earliest mention of a subsoil plough by Worlidge in 1677, yet the mind of the reader will be likely to turn at once to James Smith, of Deanstone, as the gentleman who has lately opened a very interesting and important view of tillage, by the system of subsoil ploughing. His practical knowledge has

SMITH'S SUBSOIL PLOUGH.

long been devoted to the interests of agriculture, and the results have been most beneficial. The plough he invented for the purpose is too well known to need a description, and the cut will sufficiently show its form.

RACKHEATH SUBSOIL PLOUGH.

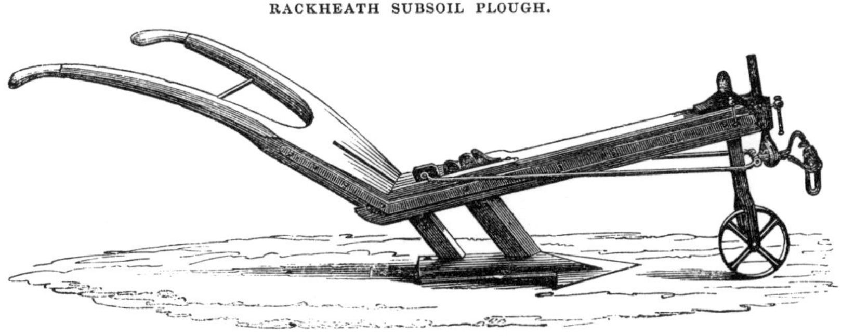

Following the Smith's Subsoil Plough, was another of a different and much lighter description, the invention of Sir Edward Stracey, Bart., Rackheath, and the plough is called by the latter name. It performs the operation of subsoil ploughing, to the depth of from ten to sixteen inches below the surface, and when preceded by the common plough, which is the plan recommended, the depth reached below the surface ground is just so much the more than the first plough effects.

RACKHEATH SUBTURF PLOUGH.

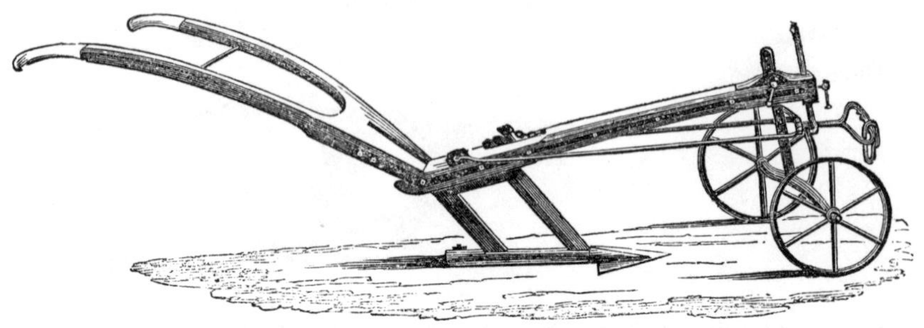

This plough is also the invention of Sir Edward Stracey. It answers admirably for under ploughing grass lands, and is made into a subturf plough by changing the wheel gear in front, to that of a carriage and two wheels, as shown below.

CHARLBURY SUBSOIL PLOUGH.

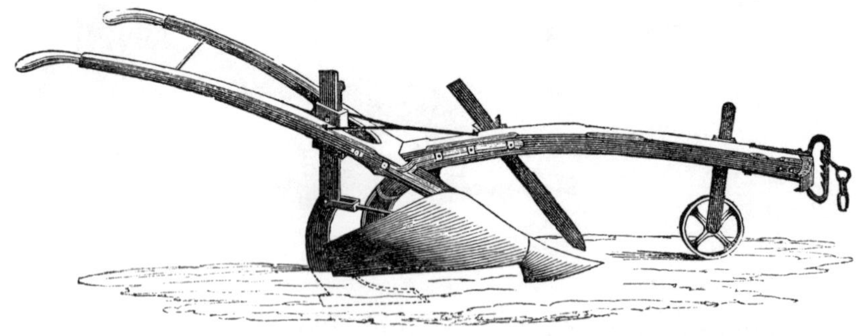

P. Pusey, Esq., in an interesting paper in *Eng. Agr. Soc. Journ.* (vol. i. p. 434), gives an account of a plough made to his order by Charles Hart, of Wantage. At the hinder part of this plough is fixed a strong tine, something like those on Biddell's scarifier, for the purpose of under-ploughing the soil. This tine is made to rise or lower at pleasure, and from the description of its use and operation given in the above paper, I should think it a valuable invention, as it may be easily attached to a plough of the common sort, as shown by the cut, and removed when not wanted.

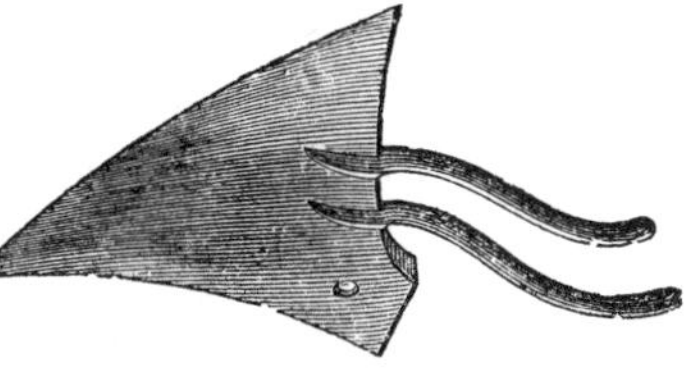

Skeleton, or cleaning ploughs, are for the purpose of loosening the land, and cleaning it from the couch-grass or rubbish which may be in it. They have no mould-boards, but simply broad shares, which are sometimes used with prongs, like the accompanying cut. The shares used are of various shapes and widths. A skeleton plough may be made by removing the mould-board from the common plough, and substituting for the share in general use, one adapted to the purpose of cleaning, as above described.

KENT TURN-REST PLOUGH.

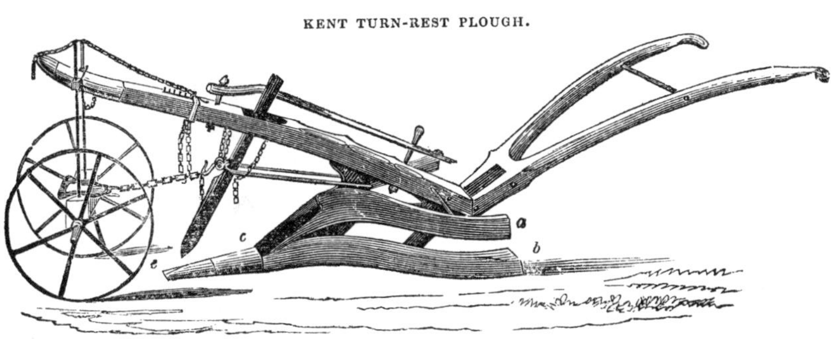

Kent turn-rest plough.—The one here shown is the plough in common use in Kent. It is intended for under surface ploughing, so as to clean it from grass and rubbish, as well as to loosen the soil. It is adapted for crossing the ridges, as well as for ploughing in a line with the common furrows, and it may be used so as to lay the "stetches" or "lands" rounding or flat as desired. There are shares to it of four, eight, twelve, and sixteen inches wide, which may be used according to the hardness or looseness of the land.

This plough, as respects its carriage and wheels, handles and beam, is the same in principle as the high gallows plough previously described; but its characteristic difference is in the construction and use of its turn-furrows, *a* and *b*, or, as they are more commonly called, its turn rests, and in its nose-piece *e*, which is called the buck. Its share *c*, is also different in shape to those on other ploughs, and it varies in width according to the work to be done with it. This plough is sometimes used without its rests as a broad-share for cleaning land, and shares are made for it from two to twenty-four inches wide.

This implement is so contrived as to lay the furrow-slices all in the same direction, from one side of the field to the other. To accomplish this, it is changed at each turn of the horses, when set in for a fresh furrow, to a right-handed or left-handed plough. The operation is as follows: the ploughman fixes the lower rest, *b*, on that side of the plough which is to turn the furrow, and he fixes the coulter on the opposite side, which is done instantly, by a wood set-stick upon the beam, as shown in the cut, called the "rod bat." Suppose a plough to be turning the furrow-slice over to the right hand, the lower rest being set so as to form the mould-board on that side, and the coulter on the left hand side as in ordinary ploughing. The soil is penetrated by the share, which is shaped like a chisel, and centred on the plough; and this is followed by the buck, which, as a wedge, raises the furrow-slice, and sets it on its edge; the upper rest, *a*, which forms the lateral wedge, then comes into operation, and finishes the turning.

As the furrow-slice requires to be completely turned over, the first furrow in the field must be sufficiently wide for the slice as it turns, first to stand on its edge, and then to fall into its place. If the furrows be required to be ten inches wide and six inches deep, there must be sixteen inches clear before the regular furrows commence and there will always be the width of one furrow left as an open one at the finish of the ploughing. This rule will apply to any width and depth in relative proportion.

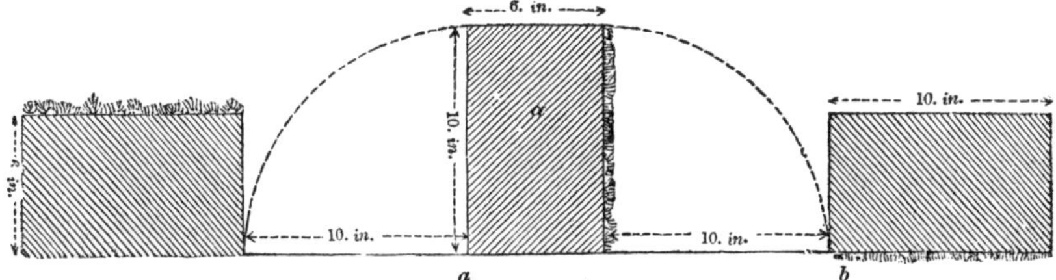

The space from *a* to *b* is first opened to the width of sixteen inches; the slice *a* then falls edgeways, into the position in which it is shown by the above figure, and is next turned completely over, so as to bury entirely the part which was its upper surface, and it remains thus turned over in the ten-inch space assigned for it, and the furrows which follow are continued in succession. It should be observed that the upper rest of the Kent plough rises as it recedes from the fore part of the plough, and the hinder part gently sweeps over the furrow slice as it rolls, thus completing the overturn. The lower rest aids the operation by pressure against the furrow-slice.

The furrow being completed to the right hand, the ploughman next changes the position of the coulter and lower rest, and proceeds as before with the adjoining furrow, but turning it to the left hand.

The Kent plough, notwithstanding its awkward appearance, great bulk, and clumsy combinations, is, in the hands of a skilful ploughman, a very effective implement, and I have frequently seen ploughing to the depth of eight inches, and with furrows eleven inches wide, so accurately performed as to astonish judges who had been accustomed to the lighter ploughs of other counties. It is, however, difficult to manage, and requires much practice to use it with any degree of facility. Having to perform both right and left ploughing, it must be perfectly true in all its bearings, and the coulter must also be so accurately fitted to an angular hole, that when the upper part is pressed either on the right or left by the rod-bat, its point shall be equally true to the edge of the furrow it is intended to cut. It is tightly braced in all its parts, the upper part of the gallows being confined by a chain to the top of the beam, and the draught chain is again braced tightly to the beam by means of a collar chain, or rather a pair of collar chains, having what is technically termed a "a goose neck" passing through one of its links, which is made circular for its admission. In order to correct any tendency to swerve from the perfect straight line, arising from this complication of braces, the ploughman is provided with a parcel of nails, of every sort and size, which he frequently carries in the foot of an old shoe nailed upon the handle of his plough; and thus equipped, should a link not be sufficiently tight, or should any tendency to "run away from its work" appear; after duly selecting a nail of the exact size fitted for the place, it is ingeniously inserted at the junction of the links, and every subsequent alteration is effected by the never-failing application to the nail store; it thus frequently happens, that by the time the plough is fitted for its work, almost every link is garnished with nails; and upon counting these on a number of ploughs at work in a trial field, the quantity of nails thus used in three, was respectively 15, 17, and 19.

As has been before observed, the work performed is nevertheless excellent, and for deep and heavy ploughing the principle is better adapted than a casual observer would suppose; but it is not to be denied that it is a more cumbrous implement than a plough formed as a turn-rest needs to be, for a large proportion of its present size and strength is requisite to provide against the strains to which it is subjected from the attempt to counterbalance the conflicting forces its erroneous construction has engendered.

The plan of laying furrows in one direction, so as to have neither ridge nor water furrows, has of late attracted more than common attention, and it has led to a careful inquiry into the system of ploughing pursued in Kent, and there seems to be a desire among many agriculturists in other counties than Kent to try the plan, provided lighter implements can be furnished for the purpose, not exceeding the power of two horses' draught. To this object some eminent practical farmers have turned their attention, and the construction of a plough, made under the direction of William Smart, a farmer of great respectability and experience at Rainham, in Kent, bids fair to open a new and very important view of the mechanical principles of the turn-rest plough. He has so successfully remodeled one on that principle, that it may be made equally applicable to the power of two or four horses, according to the work it may be required to perform. A description of this, in a letter which is published in the "Farmer's Magazine, vol. xiii. p. 59," gives a clear elucidation of his view of the subject.

This gentleman, after many trials, arrived at the conclusion, that, inasmuch as the work of the turn-rest plough depended on its wedge-like construction, its form could only be correct, in proportion to its approach to the perfect wedge ; and this form, obtained by straight lines in the direction, firstly, from the point of the share, to the throat of the plough, to produce the effect of elevating the furrow slice; and, secondly, from the edge of the coulter to the heel of the rest, to effect the turning of the flag, is that which he has adopted ; making these lines tend to an angle of 15 degrees. With the assistance of an ingenious plough-wright in his own neighbourhood, several ploughs on this principle were constructed, and these have been the basis upon which still further improvements in the detail have been carried out. Smart's plough is so constructed, that the ploughman can readily shift the coulter, and change the rest from side to side, without leaving his station behind the plough. He is enabled to effect this alteration by means of a lever, one end of which is near to the lower parts of the handles of the plough, and the other at the coulter ; and by another easy mechanical contrivance the rest is made to slip from side to side, so that each becomes alternately land-side or mould-board, as the furrow has to be turned to the right or the left. The necessity for the choice selection from the store of nails, is removed by the introduction of a screw-link and swivel in each chain. The implement constructed on the principle described, admits of being made much lighter; and its form being mechanically correct its work is effected with as little expense of draught as any single mould-board plough would require to produce a similar effect.

SMART'S IMPROVED TURN-REST PLOUGH.

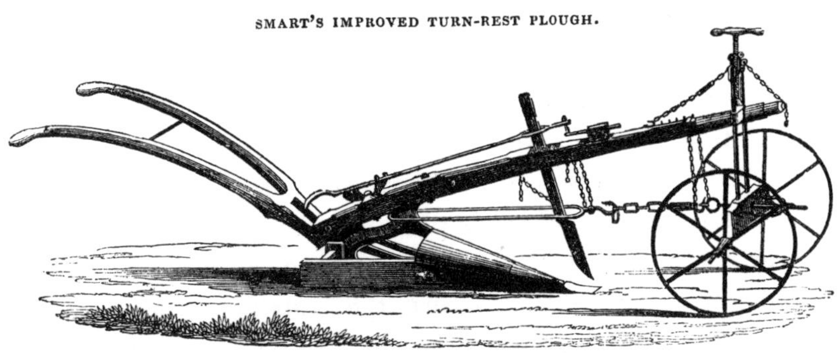

Ploughs for the purpose of turning the furrows all in one direction, and laying the slices at an angle with the horizon, as is done by the common plough, have been within the last year or two brought before the public, but none have obtained general use.

The first of these is one most ingeniously contrived by James Smith, of Deanstone. This plough has handles, beam, and frame-work, very similar to the wrought iron Scotch plough. A spindle, fixed over the beam, has attached to it a right and a left hand mould-board and share, each so fastened, that when one is in work, the other is elevated over the beam, and by turning a handle attached to a spindle, which is brought between the two handles of the plough, one fourth of a circle, the one mould-board is taken out of, and the other set ready for work. The other end of the spindle to which these are affixed, has an *eccentric*, an ingenious invention of James Wilkie, of Uddington, which, acting upon the coulter, sets it to the proper angle for the land-side of the furrow, whether right or left.

The second is the invention of Captain Hay, of Belton. This plough may be described as having a right and a left handed body, placed end to end, the beam and handles being so made as to turn on a pivot at the centre of the top of the body, so that the plough may be easily reversed, and the furrow turned to the right or left. In this case, no alteration is required in the coulter, which is fixed by wedges to the beam in the usual manner.

The third is Huckvale's plough, which is so constructed, that, by reversing the position of one of its handles, the ploughman is enabled to turn the body part from right to left, so that the part which was in the one instance the slade or sole of the plough, will alternately become its land side, and thus act on either side of the plough ; that side which is not at work forming a close cover over the other. The share is formed with two blades or cutting edges at right angles, one of which acts horizontally as a share, and the other vertically as a coulter, and the position of which is changed at each end of the furrow by the same operation. I have seen this plough used on light land and with a shallow furrow, when it appeared to do its work easily, and the position of its body was changed without difficulty.

In addition to the ploughs already described, there are two or three others which require some notice.

The Mole plough—so called from its making a hole or tunnel under ground. It is used on stony clays, for the purpose of effecting drainage. The part called the mole, that forms the tunnel, is a round iron sole, which is drawn through the land at a given depth. It varies in diameter from two to three inches, and is usually about sixteen inches long, with a point tapering five or six inches. This is attached to a strong iron shank, with a sharp edge, which passes through a mortice hole in the beam, where it is set fast by strong bolts at the requisite length, from which it admits of easy alteration. The handles are the same as those on common ploughs, but the beam and draught-iron require to be very strong, as the force required to draw this implement is very great. The usual mode is by a portable capstan and chain, with a horse walking round.

Draining ploughs—have been constructed of various forms; some on similar principles to the common plough, but much larger and deeper; others having a series of cutters to divide the soil at different and increasing depths, and with a chisel-pointed share turned up at the edges, leading upwards to an inclined plane, along which the sod, cut by the coulters vertically, and by the share horizontally, is gradually forced up and deposited upon the land. In some lands, which are free from stones, these implements may occasionally be used to advantage, but the great power required to use them, in addition to their expense, are barriers to their general adoption. The same mode of draught is applied as with the mole plough.

The Paring plough—is for the purpose of cutting and raising turf. One was exhibited at the Liverpool Meeting, which performed this work admirably. It raised the turf and set it on an edge as neatly as if done by hand. There were two angular shares with the wings facing each other, and just crossing the centre line, one being a little before the other; these were attached to shanks and preceded the mould-board, upon which the turf was raised in a similar manner to the flag in ordinary ploughing, and was deposited by the side of the cut, ready for use or burning. The inventor is Thomas Glover, near Leicester, and the plough, I understood, was made by Thomas Johnson, of that town.

Ploughs of various other shapes and fashions have from one imagined improvement to another multiplied to an almost endless variety. Each county appears to have its own peculiar favourite, differing in one or more respects from others. There is no doubt but the number of these varieties might be reduced; and for certain descriptions of soil, three or four good standard ploughs for general purposes might be adopted. But however much they may be in error who advocate these extended varieties, those who entertain the opinion that *one* implement may be made that shall suit all soils and all purposes, are not less so. The question as to what constitutes good ploughing is not a settled one; some advocate deep, and others shallow ploughing; some are content with the furrow set simply on edge; others advocate its being laid at the angle of forty-five degrees, while there are those who contend for its being turned over; and the men of Kent will be satisfied with nothing short of its being turned completely upside down. It is clear that men are thus apt to assume that the plan which succeeds best for their own locality is that which should be adopted by all others.

From this cause it is difficult to arrive at a satisfactory conclusion, and this state of things must continue, till we can lay aside early-formed predilections, and impartially test by satisfactory means the merits of each implement which may come before us in competition.

The accounts of public and private trials of the respective merits of ploughs have excited much interest, and it is to be hoped will prove of great advantage to agriculturists. At the same time it ought ever to be borne in mind that the opportunities for public trials are not always the best adapted to perfect experiments. Parties introducing them are excited by interest and emulation; and it sometimes happens that praise is given to an implement, because the work it *performs* is better done than by its competitors, while that superiority is occasioned solely by the skill of the workman. Whoever has been in the habit of attending annual public ploughing matches, must have observed that ploughs in competition, of opposite construction, have frequently exchanged places in the Award Prize List. These casual opportunities, therefore, do not satisfactorily determine the question as to which is the better implement, because their performances not being dependent on their construction solely, they are affected by other circumstances, which may be either adventitious, or otherwise. The merits of a plough is one question, and the merits of the ploughman another; and the consideration of the value of the plough itself, or the qualifications of the man who holds it, should be taken up distinctly. An average of several trials and tests should form the datum for judgment, and which should not be confined to a single experiment, however fairly such experiment may appear to have been carried out.

As the object of ploughing matches is to benefit the agriculturist, it is important that every thing relating to them should tend to so desirable an end: and while a value attaches to the encouraging efforts which have been made and are still making, to improve the art of ploughing, and the construction of the plough, I can but hope that the trials, under the direction of Agricultural Societies, will be sufficient in number, and so fairly made with reference to all parties concerned, that the judgments which issue from them, as shown by prizes and commendations, may be such as will best guide the farmer in the choice of his implements.

I have ventured the foregoing remarks on the object of Agricultural Societies, inasmuch as their influence tends to the production of more skilful workmen and more improved implements. I may also add that a ploughing match is a heart-cheering scene; and the excitement it affords to those who take a part in it, is pleasure indeed. The animation shown by the labourer, his wife and children, in the honest rivalry of the day; the good heart and humour of the farmer, who "speeds the plough" by the ready assistance he furnishes by his team, and other requisites; together with the lively interest exhibited by the presence of the nobility and gentry, give to the whole a charm that but few pleasures can outvie. It extends a beneficial influence to each class of the community, and leads to that unity of feeling in a common cause, which is one of the best securities for the good order of a neighbourhood, and the happiness of a country.

APPENDIX.

TABLE, BY JOHN MORTON, SHEWING THE DISTANCE TRAVELLED BY A HORSE, IN PLOUGHING OR SCARIFYING AN ACRE OF LAND: ALSO THE QUANTITY OF LAND WORKED IN A DAY, AT THE RATE OF SIXTEEN AND EIGHTEEN MILES PER DAY OF NINE HOURS. *(Johnson and Shaw's Farmer's Almanac, Vol. I. p.* 191.*)*

Breadth of Furrow-slice, or Scarifier.	Space travelled in ploughing an Acre.	Extent ploughed per day, at the rate of		Breadth of Furrow-slice, or Scarifier.	Space travelled in ploughing an Acre.	Extent ploughed per day, at the rate of	
Inches.	Miles.	18 Miles. Acres.	16 Miles. Acres.	Inches.	Miles.	18 Miles. Acres.	16 Miles. Acres.
7	$14\frac{1}{8}$	$1\frac{1}{4}$	$1\frac{1}{8}$	46	$2\frac{1}{6}$	$8\frac{1}{3}$	$7\frac{2}{5}$
8	$12\frac{1}{4}$	$1\frac{1}{2}$	$1\frac{1}{4}$	47	$2\frac{1}{10}$	8	$7\frac{3}{5}$
9	11	$1\frac{3}{5}$	$1\frac{1}{2}$	48	$2\frac{1}{12}$	$8\frac{3}{4}$	$7\frac{3}{4}$
10	$9\frac{9}{10}$	$1\frac{4}{5}$	$1\frac{3}{5}$	49	2	$8\frac{9}{10}$	$7\frac{9}{10}$
11	9	2	$1\frac{3}{4}$	50	2	$9\frac{9}{10}$	$8\frac{1}{10}$
12	$8\frac{1}{4}$	$2\frac{1}{5}$	$1\frac{9}{10}$	51	$1\frac{9}{10}$	$9\frac{1}{5}$	$8\frac{1}{4}$
13	$7\frac{1}{2}$	$2\frac{1}{3}$	$2\frac{1}{10}$	52	$1\frac{9}{10}$	$9\frac{1}{2}$	$8\frac{2}{5}$
14	7	$2\frac{1}{2}$	$2\frac{1}{4}$	53	$1\frac{9}{10}$	$9\frac{3}{4}$	$8\frac{1}{2}$
15	$6\frac{1}{2}$	$2\frac{3}{4}$	$2\frac{2}{5}$	54	$1\frac{4}{5}$	$9\frac{4}{5}$	$8\frac{9}{10}$
16	$6\frac{1}{6}$	$2\frac{9}{10}$	$2\frac{3}{5}$	55	$1\frac{4}{5}$	10	8
17	$5\frac{3}{4}$	$3\frac{1}{10}$	$2\frac{3}{4}$	56	$1\frac{3}{4}$	$10\frac{1}{4}$	9
18	$5\frac{1}{2}$	$3\frac{1}{4}$	$2\frac{9}{10}$	57	$1\frac{3}{4}$	$10\frac{2}{5}$	$9\frac{1}{5}$
19	$5\frac{1}{4}$	$3\frac{1}{2}$	$3\frac{1}{10}$	58	$1\frac{7}{10}$	$10\frac{3}{5}$	$9\frac{1}{3}$
20	$4\frac{9}{10}$	$3\frac{3}{5}$	$3\frac{1}{4}$	59	$1\frac{7}{10}$	$10\frac{3}{4}$	$9\frac{1}{2}$
21	$4\frac{7}{10}$	$3\frac{4}{5}$	$3\frac{1}{3}$	60	$1\frac{3}{5}$	$10\frac{9}{10}$	$9\frac{7}{10}$
22	$4\frac{1}{2}$	4	$3\frac{1}{2}$	61	$1\frac{3}{5}$	$11\frac{1}{5}$	$9\frac{4}{5}$
23	$4\frac{1}{4}$	$4\frac{1}{5}$	$3\frac{7}{10}$	62	$1\frac{3}{5}$	$11\frac{1}{3}$	10
24	4	$4\frac{1}{3}$	$3\frac{9}{10}$	63	$1\frac{3}{5}$	$11\frac{1}{2}$	$10\frac{1}{5}$
25	4	$4\frac{1}{2}$	4	64	$1\frac{1}{2}$	$11\frac{7}{10}$	$10\frac{1}{3}$
26	$3\frac{4}{5}$	$4\frac{3}{4}$	$4\frac{1}{5}$	65	$1\frac{1}{2}$	$11\frac{4}{5}$	$10\frac{1}{2}$
27	$3\frac{3}{5}$	$4\frac{9}{10}$	$4\frac{1}{3}$	66	$1\frac{1}{2}$	12	$10\frac{3}{5}$
28	$3\frac{1}{2}$	$5\frac{1}{8}$	$4\frac{1}{2}$	67	$1\frac{1}{2}$	$12\frac{1}{4}$	$10\frac{4}{5}$
29	$3\frac{1}{2}$	$5\frac{1}{4}$	$4\frac{3}{5}$	68	$1\frac{1}{2}$	$12\frac{2}{5}$	11
30	$3\frac{1}{3}$	$5\frac{3}{4}$	$4\frac{4}{5}$	69	$1\frac{2}{5}$	$12\frac{3}{5}$	$11\frac{1}{8}$
31	$3\frac{1}{5}$	5	5	70	$1\frac{2}{5}$	$12\frac{3}{4}$	$11\frac{1}{3}$
32	$3\frac{1}{10}$	$5\frac{4}{5}$	$5\frac{1}{4}$	71	$1\frac{2}{5}$	$12\frac{9}{10}$	$11\frac{1}{2}$
33	3	6	$5\frac{1}{3}$	72	$1\frac{2}{5}$	$13\frac{1}{8}$	$11\frac{3}{5}$
34	$2\frac{9}{10}$	$6\frac{1}{5}$	$5\frac{1}{2}$	73	$1\frac{1}{3}$	$13\frac{1}{3}$	$11\frac{4}{5}$
35	$2\frac{4}{5}$	$6\frac{1}{3}$	$5\frac{3}{5}$	74	$1\frac{1}{3}$	$13\frac{1}{2}$	12
36	$2\frac{3}{4}$	$6\frac{1}{2}$	$5\frac{4}{5}$	75	$1\frac{1}{3}$	$13\frac{3}{5}$	$12\frac{1}{8}$
37	$2\frac{2}{3}$	$6\frac{1}{4}$	6	76	$1\frac{3}{10}$	$13\frac{4}{5}$	$12\frac{1}{4}$
38	$2\frac{3}{5}$	$6\frac{9}{10}$	$6\frac{1}{8}$	77	$1\frac{3}{10}$	4	$12\frac{1}{2}$
39	$2\frac{1}{2}$	$7\frac{1}{8}$	$6\frac{1}{3}$	78	$1\frac{1}{4}$	$14\frac{1}{4}$	$12\frac{3}{5}$
40	$2\frac{1}{2}$	$7\frac{1}{3}$	$6\frac{1}{2}$	79	$1\frac{1}{4}$	$14\frac{2}{5}$	$12\frac{3}{4}$
41	$2\frac{2}{5}$	$7\frac{3}{4}$	$6\frac{3}{4}$	80	$1\frac{1}{4}$	$14\frac{3}{5}$	$12\frac{9}{10}$
42	$2\frac{1}{3}$	7	$6\frac{2}{3}$	81	$1\frac{1}{5}$	$14\frac{3}{4}$	$13\frac{1}{10}$
43	$2\frac{3}{10}$	$7\frac{4}{5}$	7	82	$1\frac{1}{5}$	15	$13\frac{1}{4}$
44	2	8	$7\frac{1}{10}$	83	$1\frac{1}{5}$	$15\frac{1}{8}$	$13\frac{2}{5}$
45	$2\frac{1}{5}$	$8\frac{1}{6}$	$7\frac{1}{4}$	84	$1\frac{1}{6}$	$15\frac{1}{3}$	$13\frac{3}{5}$

SECTION III.

THE HARROW, SCARIFIER, &c.

Next in antiquity and usefulness to the plough in the implements of agriculture, is the harrow. This instrument in some rude form or other must have existed from a very early period. For as the preparation of the ground for the reception of the seed required one kind of implement, so the covering of the seed with the loosened soil, required another of a different form. The first and most readily found harrow was, probably, merely the branch of a tree; even so late as 1668, Gervas Markham in his "Farewell to Husbandry," p. 61, gives a wood engraving of a harrow the directions for the manufacture of which he thus states, "get a pretty big white thorn tree, which we call the hawthorn tree, and make sure that it be wonderful thick, bushy, and rough grown." The natives of some parts of India, in fact, still use no other instrument. It was a further improvement to construct a wooden harrow by fastening together the branches of the fir-tree, leaving their partially removed spurs to serve as teeth. The peasants of some portions of Russia still employ such a harrow. The joining together of wooden frames without teeth, as in the *Haken* of the Belgian farmers, was a later improvement—then came the addition of wooden teeth, next the use of iron teeth with wooden frames, and, lastly, the construction of most descriptions of harrows entirely of iron.

This instrument, indeed, succeeds to the plough in the natural order of description, and in the uses to which it is applicable. Its purposes are to pulverise the ground which has been moved by the plough, to disengage from it the weeds and roots which it may contain, or to cover the seeds of the cultivated plants, when sown. The form of the plough, as before shewn, has been very different in different ages and countries, and there is little resemblance between the rude machines of the ancients and some of those which are now employed; but the harrow seems to have been nearly of the same form from the earliest times to which we are able to trace it on sculptures, medals, and other remains of antiquity. It is, in fact, a much more simple machine than the plough; yet it is an instrument of great utility in tillage, and, with one exception hereafter noticed, no other has yet been devised to supersede its use, or to equal it, for many of the purposes to which it is applicable. (*Quart. Journ. of Agr.* vol. i. p. 503.)

Important as is the operation of harrowing, and second only to that of ploughing, it has often appeared to me that these implements have scarcely obtained the attention which is their due. I here speak less with reference to the improvements which have been carried into effect, than to the selection which appears generally to have been made. The operation is in many neighbourhoods so performed as to exhibit a prominent defect, either in the management of the farm, or in the construction of the implement. Perhaps the blame may here be fairly shared. It is admitted by all acquainted with the subject, that harrowing, especially on heavy soils, is the most laborious operation on the farm,—not so much, perhaps, on account of the quantum of power requisite for the draught (though this is sometimes considerable), as for the speed with which the operation is, or ought to be, accompanied; and yet it is frequently left to the charge of mere boys, and sometimes performed by the worst horses on the farm.

If we examine a field, one half of which has been harrowed with weak inefficient horses, and whose pace was consequently sluggish, the other half with an adequate strength and swiftness of animal power, we shall find the former will be rough and unfinished; the latter comparatively firm and level, and completed in what would be called a husbandry-like manner. Scarcely any thing in farming is more unsightly than the wavy serpentine traces of inefficient harrowing. The generality of harrows appear too heavy and clumsy to admit of that dispatch without which the work cannot be well done; and though it is evident that different soils demand implements of proportionate weight and power, yet, for the most part, harrows have been rather over than under weighted, particularly when employed after a drill, or to bury seeds of any kind.

Harrowing has been so long regarded as an operation which must be attended with considerable horse-labour, that attention does not appear to have been sufficiently turned to the inquiry whether this labour might not be greatly reduced by lightening the instruments with which it is performed. Many would be surprised at the amount of reduction of which seed harrows, at least, are capable, and where land is clean, to see how effectively a gang of very light small-toothed harrows may be used.

Having noticed, in some parts of Norfolk, the perfect manner in which seed corn is covered by a common rake with wooden teeth, a friend of mine constructed a gang of harrows on the following plan, and he states that it proved the most popular and useful implement of the kind to the farm.

GANG OF LIGHT SEED HARROWS.

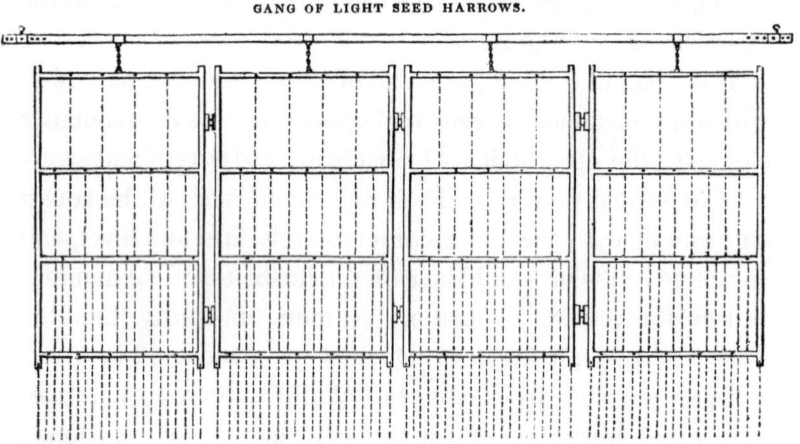

The frames are of ash, and as light as possible, with iron teeth being but three inches long, exclusive of the part which enters the wood-work. They screw into the balks in the manner shown in the annexed figure.

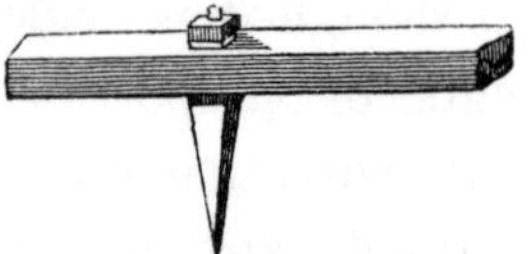

It should be observed that the above four harrows are amply sufficient to cover a twelve-furrow stetch or ridge of 108 inches, but three will be wide enough for a three-furrow stetch of 90 inches, exclusive of a small portion of the furrows. If for some purposes the teeth be found too thick, every alternate tooth may be taken out ; but for general purposes this will hardly be necessary. The two horses require, on this plan, to be kept quite level, for if one be suffered to go in advance of the other, a diagonal line is produced, by which the teeth will be made to follow each other, instead of cutting fresh ground. I am aware that, by the usual construction of harrows, a diagonal line of draught is required, in order to throw the teeth into a proper working position; but I am strongly inclined to the opinion, that the correct working of the implement ought to depend on its construction, and not on any particular mode of working it. Besides, the system of keeping one horse in advance of his partner is bad in principle; it is an unequal division of labour, the fore-horse being compelled to do more than his share of the work, which, under any circumstances, is always heavy enough. The balks of the above set sf harrows were made of wood in order to insure extraordinary lightness; but, for general purposes, I prefer those made of iron, the weight of which can be increased to any reasonable degree without adding much to their substance. This is important in working tenacious clays, which, by adhering to the common clumsy wooden balks, considerably increase the labour, and at the same time impede the proper execution.

In an experiment made between a pair of wooden harrows and a pair of iron ones, constructed on the same plan, having the same number, and precisely the same disposition of the teeth and frames, although those of iron were found to be 20 lbs. lighter than those of wood, yet the former worked decidedly better and steadier than the latter; in fact, the iron harrows cut into the land, while those made of wood rode, or rather danced, on the surface. Next, as to the length and position of harrow teeth. The common plan is to set them springing a little forward, and gradually increasing in length from the fore to the hind row. There is no advantage in this, but the contrary; for, if the action of harrows so constructed be carefully examined, it will be found the reverse of what it ought to be,—the hind part will be thrown up, and the fore teeth, short as they are, will have to do all the work. In some experiments made with harrows, the fallacy of the idea, that an inequality in the length of the teeth was essential to the proper working of harrows, was made evident. For this purpose, a harrow constructed on the old-fashioned plan of unequal and springing teeth, was reversed, putting the longest teeth in front; the whole of the teeth then pointing *backwards* instead of *forwards*. Nothing could work better; there were no chucks and snatches, but all went on smoothly and steadily. I do not, from this circumstance, recommend harrows to be so constructed, but have no doubt that each harrow should have all its teeth of equal length, and they should stand perpendicularly from the frames.

GANG OF HEAVY IRON HARROWS.

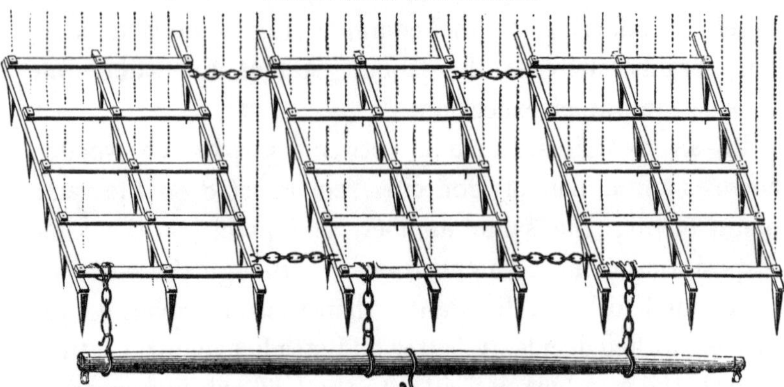

The above engraving of iron harrows is introduced to show the form in which they are usually made; they are used in gangs of three, four, or five, as may be required to suit the lands on which they are used, and may be made to any weight required.

Armstrong's Harrows.—These instruments differ from others in the form of their balks or framings, which are of iron, and of a zigzag shape, so arranged that the tooth or tine shall be fixed at each angle, in such manner that the lines formed by them shall be equidistant over the breadth of the land they are intended to cover; by this arrangement the tines are placed at greater distance from each other, and thus on foul land, or in wet weather, they are less liable to clog.

The inventor claims, as a prominent advantage in these harrows—the arrangement for the draught; this, it will be seen by the sketch, is obtained from a hake in the centre of the whipple-tree, or bar, to which the harrows are attached, so that, should the horses draw unequally, no disadvantage results from throwing the harrows out of the straight line, and the teeth consequently out of track, an accident to which, in the common mode of draught, harrows are constantly liable.

ARMSTRONG'S PATENT HARROWS.

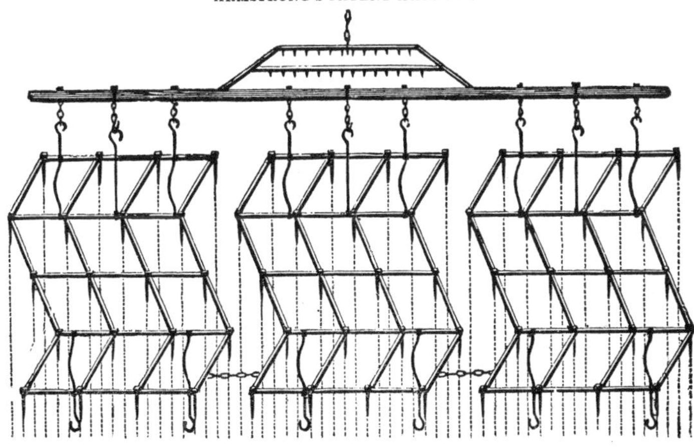

Armstrong's harrows are made by J. Howard, of Bedford, and can be adapted either for heavy or light work.

At the R. A. S. E. meeting, at Bristol, 1842, a gang of harrows was exhibited, the invention of Evan W. David, of Radyr Court, and obtained the Society's prize; of the working these, the writer has had no opportunity of forming a judgment, but unites in the opinion expressed in the report of the judges at that meeting, "that much ingenuity was displayed in the mode of combining several harrows together so as to accommodate them to ridges, and to cover a large breadth of land," and that they exhibited an unusual combination of lightness, strength, and good workmanship. These were manufactured by Saunders and Williams of Bedford.

At the same meeting James Smith, of Deanston, exhibited three varieties of an implement which he terms a *Chain-brush* or *Web-harrow*, adapted either for light or heavy operations in pulverizing soil; these are constructed by arranging in lines a series of iron rings, or rather discs, with a hole sufficiently large in the centre of each, so as to allow of their revolving freely, or rolling from side to side, around the rods of the loose iron web by which they are held together; a familiar notion may be obtained of their mode of operation by imagining six or eight quoits with a chain threaded through their centres, and behind them a similar set working in the intervals left by the first row, and so on for a sufficient number of rows to form a square. It will then be seen, that the effect produced by the weight of the web formed by the quoits and chains as it is drawn over the surface will be to press the edges of the discs into the soil, and, as they severally revolve, they cannot fail to sever the clods with which they may be brought into contact.

I now proceed to give a brief description of some other implements intended to effect the same purposes as the common harrow, but of a more elaborate character.

Morton's Revolving Brake-Harrow.—This is an ingenious, and, on light sandy soils, a very effective implement. The principle is somewhat similar to that of the haymaking machine, (*See* RAKES); except that in place of the surface it goes to the very bottom of the furrows, bringing up a far greater quantity of weeds than any fixed harrow could be expected to accomplish. One of these implements was some years ago in use on the Earl of Leicester's estate at Holkham, and many were struck with its singular capabilities for clearing and pulverizing the soil. It seemed admirably calculated for its intended purpose; and I have no doubt that the quantity of rubbish it brings to the surface must supersede the necessity of an extra ploughing; still, with all its merits, it does not appear to have made much way in England, for, with the above exception, I do not remember to have seen one at work. I have recently, however, had the opportunity of witnessing the performance of a revolving harrow, of a somewhat similar construction, but with considerable improvements in its details; this is the invention of a gentleman of the name of Brite, residing at Teddesley Park farm, in Staffordshire, to whom I am indebted for providing me ample opportunity for seeing the implement in operation; in one portion of the land large stones abounded to an extent that would have impeded, if not prevented, the working of any common instrument, but with the power of four horses, it performed its work admirably, and I believe it would be found a very useful implement on many soils where weeds abound.

VAUX'S PATENT REVOLVING HARROW.

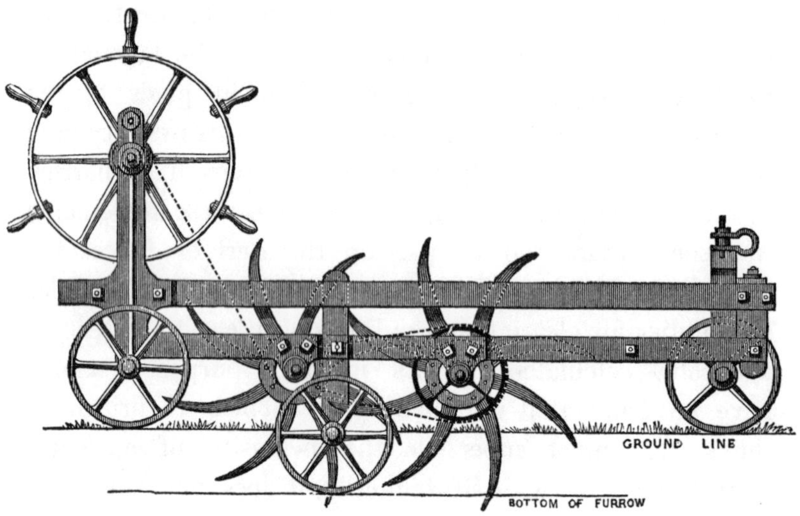

Vaux's patent Revolving Harrow for Agricultural Purposes.—Of this implement a sketch is subjoined, taken from the specification of the patentee obtained in 1836, by which its construction and mode of working will be clearly seen. It has this important feature, that one part of the apparatus, by its rotatory motion, serves to clear the other part; but, unless the harrows were of larger circumference than they appear to be, I fear it would have great tendency to choke on wet land.

Biddell's Extirpating Harrow.—This is a new implement, invented by Arthur Biddell of Playford, and similar to the scarifier which bears his name. It is intended for breaking up land when it is too hard for the heaviest harrows, and for bringing winter fallows into a state of fine tillage. In working summer-lands it is calculated, by the shape of its teeth, to bring to the surface all grass and rubbish. The teeth are placed in three rows, in order to allow sufficient distance from each other to prevent choking, and the implement is so constructed as that, by means of levers, the teeth may be elevated or depressed at pleasure. According to the form of the lands it may be required to operate upon, it may either be used perfectly parallel, or the fore-teeth may be made to penetrate deeper than the hinder ones, whilst those at either side may, when one wheel is required to run in the furrow, be instantly adjusted to the level of the land, so that every tine shall penetrate to an uniform depth of six inches if required ; and they will work equally well at any less depth.

BIDDELL'S EXTIRPATING HARROW.

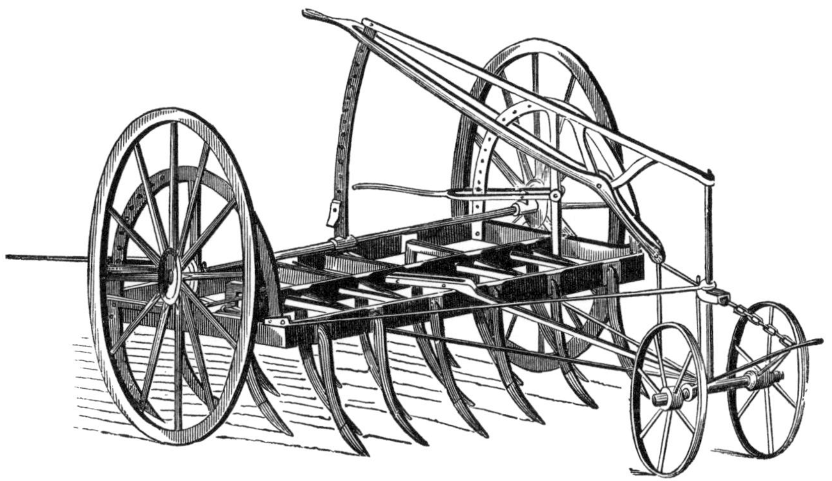

I have frequently seen this implement at work on very foul land and on stubbles, when it has been too hard to allow the use of the plough. As the interval between the lines formed by its teeth does not exceed four inches, the soil has been completely stirred. The tines may be either used with points or with steel hoes; and with the latter the skimming, or, as it is frequently called, the "broad-share" process, may be quickly accomplished. The weight is not found to be a disadvantage, but, from the stability it gives, the contrary; and being borne on high wheels it does not require so much horse-labour as might be supposed, two horses on most soils being generally sufficient.

SECTION IV.

THE SCARIFIER.

With the harrows are intimately connected the scarifiers. They approach so closely in their character that it is not easy to draw the line of distinction, indeed the last mentioned implements may, with equal propriety, be classed with the scarifiers. These latter were the production of an age of considerable agricultural advancement. It was only after the poorer soils had, by the necessities of mankind, been forced into cultivation, and the less valuable lands had become infested with the couch-grass, and other weeds, which tenant the long cultivated poorer descriptions of arable land, that the necessity was generally felt of a more powerful implement for bringing to the surface the roots of the stubborn weeds thus growing, hence arose the first attempted variation from the form and principle of the common harrow, and finally the invention of the now highly useful class of scarifiers, scufflers, and extirpators.

Among the earliest of the many varieties of this implement was one which the late T. Cooke used, attached to the frame-work of his drill, the coulters and apparatus of which, being removed, gave place to a bar, or head, suspended by joints to the axle, on which a row of strong tines was fixed.

On a similar plan to this, but working on a plough carriage, another invention by Robert Fuller, a practical farmer of Ipswich, about thirty years ago, came into operation, and worked exceedingly well. Its character will be seen from the following cut; the tines are suspended upon a round axle, and may be turned up out of work at pleasure when fewer are required, or at wider intervals.

FULLER'S EXTIRPATOR.

This implement is still in use in the county of Suffolk, but it is now generally constructed with two beams and a double row of tines; by this alteration double the space is allowed between each tine, the implement is less liable to choke, and works more freely through the soil. The carriage upon which the beam rests is similar to the common wheel-plough carriage, or gallows, described at p. 33.

FINLAYSON'S PATENT HARROW.

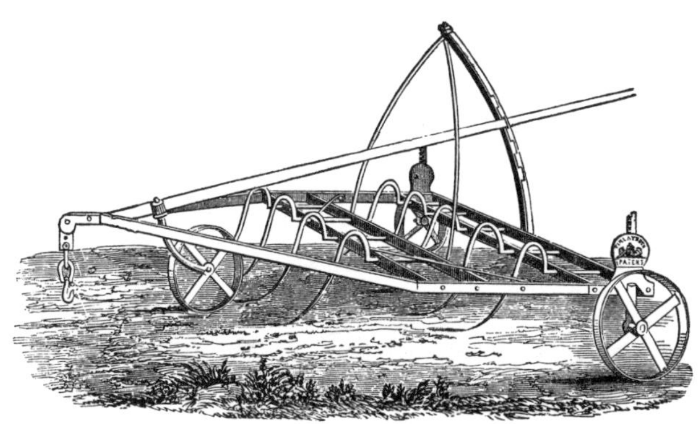

A great improvement on Fuller's Extirpator was *Finlayson's patent self-cleaning Harrow* as shown in the above sketch. This well-known implement may be called the parent of several of the same description, which, in improved forms, have subsequently come into use.

WILKIE'S PARALLEL ADJUSTING BRAKE.

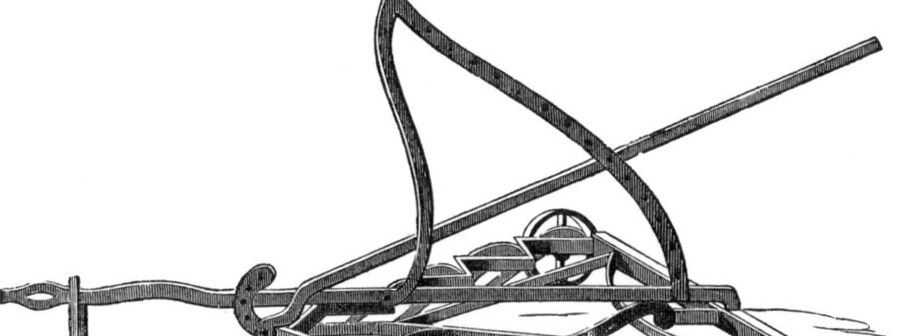

Wilkie's parallel adjusting Brake.—This is very nearly allied to the implement last mentioned: its chief improvement consists in the triangular adjustment of the teeth or prongs, and the facility with which they may be completely thrown out of work; whereas, with Finlayson's harrow this can only be partially done, the hind teeth of the latter still retaining some hold of the ground, even though the first row be lifted up. This, I am aware, has been represented as an advantage, inasmuch as the slight hold retained by the back row of tines prevents the implement from running on the horse's heels, when turning at the ends of the stetches on hilly ground. There is but little in this as an argument in favour of any implement of the kind. I am disposed to give the preference to one like Wilkie's brake, which, by a parallel movement of the frame in which the tines are fixed, can, either at the turnings or while in action, be elevated or depressed sufficiently to clear itself.

KIRKWOOD'S GRUBBER.

Kirkwood's Grubber, in its operation, somewhat resembles those last described; but is superior to them in working. The leverage that is obtained by pressing on the handles or stilts of the machine, whether in action or at rest, is so simple and yet so powerful in its effect as to regulate the depth of the tines to the greatest nicety; or, in cases of obstruction, to throw them out of work altogether. It is an admirable implement, and well deserving the high commendation which is generally bestowed on it, although inferior in the form of its teeth to either of the before described implements.

BIDDELL'S SCARIFIER.

Biddell's Scarifier.—This implement, of which the above engraving will give a general outline, is the invention of Arthur Biddell, of Playford, and was first made under his direction about thirty years ago, with a framing of wood and tines of wrought iron. It is only within the last few years that it has been made principally of cast iron and generally introduced.

It consists of a very strong cast iron frame, upon which the teeth, nine in number, are arranged in two rows, and so disposed at intervals of sixteen-and-a-half inches from each other, as that those in the hinder row should form a path midway in the intervals left by the passage of those in the front. It is suspended on a cranked axle between two wheels, fifty inches in height behind, and on an upright shaft, carried on two small wheels running close together in front. It will be seen that the machine is thus suspended on three points, and by means of two levers, the one to direct the position of the front teeth, and the other to regulate the depth of the hinder ones; it may be balanced between these points in any direction that may be required.

It may thus be used either with the fore tines parallel with the hinder ones, or at a greater or less depression; this arrangement allows it to penetrate very strong or hard land, and to retain its hold when scarcely any other implement would produce any effect, and even when the plough could not work to advantage. By a simple contrivance to shift the bearing of the frame upon the axle, either side of the machine may be depressed so that the tines shall penetrate the land to a uniform depth, even when, from the circumstance of one wheel having its path along the furrow, the bottom of the wheels may not be parallel with the general level of the land. The tines are prepared to fit case-hardened cast iron points, of one, two, or three inches width, or cast iron or steel hoes of nine inches width; with these latter every part of the land will be cut. They are readily taken off and exchanged.

Having thus generally described the Biddell's scarifier, I now proceed to lay before my readers a description of another implement since introduced as Lord Ducie's Cultivator; and as in the Prize Essay on Agricultural Mechanics, recently printed in the Journal of the Royal Agricultural Society of England, a detailed description is given by its author, the talented manager of Lord Ducie's example farm, I quote his own description as the fairest mode of stating its advantages.

" The Uley Cultivator rests on four wheels, the front ones being eighteen inches in diameter, and the hind ones three feet four inches in diameter. From the front ones rises an iron rod, having a circular section which passes through the point or nose of the framework; the hindmost ones rest on cranks on an axle which runs straight through the frame-work: the frame-work is thus twenty inches off the ground."

" 2nd. The teeth are not placed in rows, but are arranged as

LORD DUCIE'S CULTIVATOR.

in the annexed figure, and somewhat similar to the arrangement in Kirkwood's grubber. The space between each is thus two feet—twice that of Scoular's—while the distance between the paths of each is only six inches, the same as in Scoular's. On the points of these teeth, either chisel-shaped or duckfoot or triangular edges for paring are slipped, the form of the tooth on which they slip being such as to

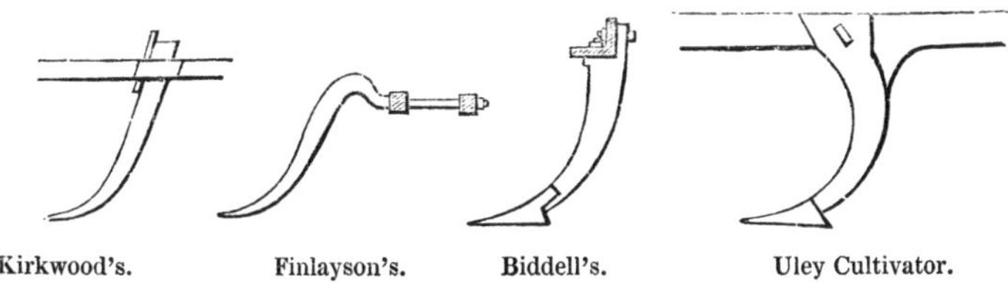

Kirkwood's. Finlayson's. Biddell's. Uley Cultivator.

hinder these teeth from slipping round. This, with the form of the teeth, is represented in the figure; the teeth for paring are much more pointed than those of Biddell's, with which they are compared in the next figure. The implement has been tried for paring as well as for stirring, and is found to move the whole surface of the ground most perfectly."

"3rd. The frame-work does not, as in Finlayson's harrow, consist of a series of cross bars, into two of which the teeth are fixed in two rows, but it is as in the annexed figure; each tooth having its own bar, to which it is affixed, and not being attached in a direction across the bar, but in the direction of its length, so that any strain on it cannot tend to twist it."

"4th. The tooth is not of the same form as Finlayson's, but it has, like Finlayson's, a self-cleaning form. This will be understood from the figure of it given above. It will be seen that the rise for the first five inches is most gradual, and that any couchy earth or clods coming against it will not be stopped by it and clog it, but will rise and tumble over, and fall aside, and so pass by. This, with the height of the frame mentioned in No. 1, and the distance between the teeth, spoken of in No. 2, hinders any possibility of the machine being clogged in fast land."

"5th. The machine is partly cast iron and partly wrought; and thus, while the liability of Biddell's to break is avoided, the advantage of having the cheapness of cast metal is partly ensured. The sections of the teeth, given in a figure above, show that they are strong enough. Their strength is increased by having an abutment on the frame, against which they rest. The manner in which they are fixed to the frame is also such as to increase their strength. They are made slightly taper at the place where they are keyed in, so that a blow of the hammer makes them fit perfectly accurately to the frame. They are not fixed by wedges, as the coulter of a plough, but by a key which passes through the frame, through both the mortice and tenon, and thus holds them fast."

"6th. The mode in which it is raised out of the ground, and the plan by which its depth is regulated, is an invention of Mr. Clyburn, of Uley Works. The ease by which the operation is performed is greater, and the complication of Biddell's scarifier is avoided, while the regularity and parallelism of the motion of the frame-work, as it is raised or lowered, is as perfect as in Scoular's or Kirkwood's."

"The hinder part of the machine is, as in Scoular's implement, raised by increasing the inclination of the cranks which attach the hind wheels to the frame. In this, however, instead of inclining backwards, and thus requiring, on lifting the machine, an exertion of force, as we before explained, against the whole pull of the horses, the crank is thrown forward exactly the other way; and thus the whole force of the horses, instead of being exerted against the effort of the driver to raise the implement out of the ground, is actually exerted in assisting him to do so."

"The front part of the machine is not raised in the same way as the hinder part, by a crank on the front wheel, but it slips up and down the vertical pole attached to them, and is raised in the following way: the cranks of the hinder wheels are attached to an axle, which passes through the frame, and thus supports it. On this axle a toothed wheel is fixed: this toothed wheel is worked by an endless screw on a shaft, which also works in the frame of the machine. This shaft working by means of this endless screw, a toothed wheel on the axle across the implement is necessarily parallel to the length of the machine, and is turned by means of a crank-handle above. Turning this handle round in one direction, it is evident that the axle on which the toothed wheel is fixed will be turned round, and the crank of the hind wheels raised, and thus the frame brought nearer the surface of the ground, the teeth let deeper into it; by turning the handle the other way, the hind part of the machine is raised."

"In order to get the front part of the machine raised and lowered exactly as the hinder part of it, the following simple method is adopted: on the axle of the hinder wheels a crank is fixed, attached for greater simplicity to the pinion before mentioned. It is the same length as the crank of the hind wheels; when, therefore, the pinion is turned round, these wheels being by this crank raised, that is, the frame being lowered, say two inches, the extremity of this crank is lowered four inches.* From the crank a vertical rod rises, which is attached to one end of a lever, resting at the other end on the vertical rod at the front of the machine, and attached exactly at the middle by means of a rod of iron to the frame-work of the implement. When, therefore, the point of the crank is lowered four inches, the hindmost extremity of the lever is also lowered four inches, and, as it rests on the top of the vertical rod at the other end it depresses the rod in the middle and the frame-work attached to it exactly two inches, the same as the hind part of it. The perfect parallelism of its motion is thus ensured, while the mode of lifting it requires but a small exer-

* Two inches by the action of the crank, and other two inches by the lowering of the frame to which the crank is attached.

tion of force. The implement is altogether about six cwt. in weight. It is supplied at Uley, at prices varying according to the size, weight, &c., at which it is ordered, certain forms of the implement being intended for two horses, and others for four."

In the Prize Essay from which the foregoing description is quoted, a comparison is instituted between this implement and its predecessor, less favourable to the Biddell scarifier than, I think, would have been the case had the author been more fully acquainted with the construction and uses of the original implement than appears to be from his description of it, in which occur the following statements :—

1st. That it is a clumsy implement.

2nd. That the apparatus for adjusting the tines is complicated.

3rd. That the form of the tine has nothing to recommend it.

4th. That it is made of cast iron, and, therefore, liable to break.

5th. That four horses are required to work it when made at a width of 4ft. 6in.

Taking these as the points upon which the superiority of Lord Ducie's implement is assumed—let us examine the two implements in detail, with reference to each, and we shall be better able to form a correct judgment of their comparative advantages.

1st. As to comparative clumsiness.

Biddell's scarifier, having nine tines, covering a breadth of 5ft. 6½ in. weighs 9 cwt 2 qrs. 5 lbs., equivalent to about 15¼ lbs. for every inch of width.

Lord Ducie's cultivator, having five tines, covering a breadth of 2ft. 11½in.; weighs 7cwt. 19lbs. equivalent to about 23 lbs. for every inch of width.

2nd. Comparative simplicity of apparatus for adjustment.

In Biddell's scarifier the adjustment is accomplished by two levers, one of which adjusts the frame or body of the implement, and the other enables the user to elevate or depress either the front or hinder teeth as he may require, The lever regulating the general depth will not require to be moved from the time of commencing the operation till it finishes it, therefore the only movement required while the implement is at work, or when turning at the ends, is obtained by the simple elevation or depression of one lever.

The apparatus for accomplishing the adjustment of Lord Ducie's cultivator is fully described at p. 83, and through the four following paragraphs, and the motion is confined solely to a parallel one.

3rd. Comparative form of tines.

In the Essay this comparison is made upon an assumed figure for the Biddell's tine, which, had it been correct, would have been worthy the condemnation the author has bestowed upon it. As the illustration he has given of it, see p. 79, does not furnish a correct idea of its form, I here give cuts, made from working drawings of the tines of each implement. Scale ¾ inch to a foot.

BIDDELL'S TINE.

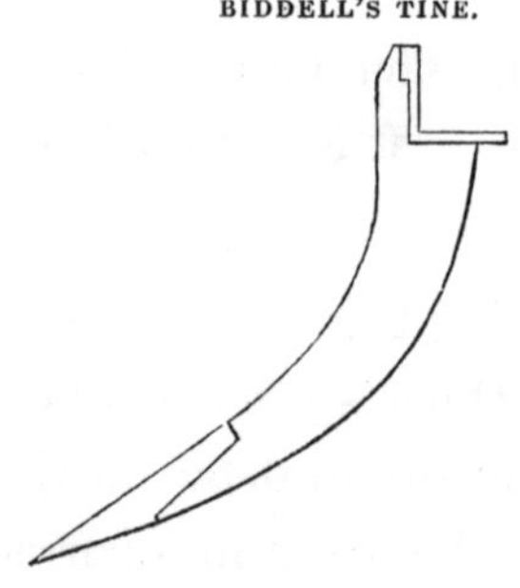

The advantage claimed for Biddell's form of tine is that, while its shape is that which will penetrate the soil under all circumstances, its curvature is, at the same time, calculated to raise the rubbish to the surface without breaking it, the necessary strength is secured in the line of the greatest resistance, while the thin edge which has to pass through the land is retained.

LORD DUCIE'S TINE.

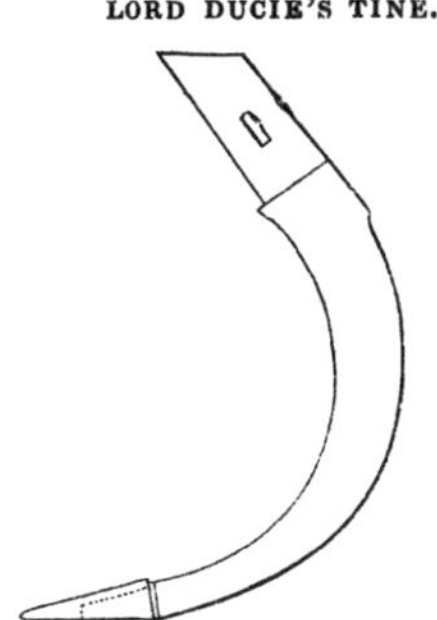

4th. Comparative adaptation of the material of which each implement is constructed.

Biddell's scarifier is constructed of a cast-iron frame, cast-iron tines, wrought-iron braces, lever, and axle; wheels of wood, with hoop tires.

Lord Ducie's is constructed of a cast-iron frame, cast-iron tines, cast-iron wheels, cast-iron worm-wheel, wrought-iron lever and axle, no wrought-iron braces.

5th. Comparative power required by the two machines.

The following is the result obtained and reported by the judges of the Richmond Agricultural Society, after a long and patient trial.—*Vide Mark Lane Express*, September, 1842.

Biddell's scarifier, having nine tines with 1 in. chisel points, working 3½ in. deep, and measuring 5 ft. 6½ in. in width, average draught 6 cwt., being equivalent to $10\frac{1}{10}$ lbs. for every inch of width.

Earl Ducie's cultivator, having five tines with 2 in. points, working 3½ in. deep, and measuring 2 ft. 11 in. in width, average draught 6 cwt. 0 qrs. 7 lbs., being equivalent to $19\frac{2}{5}$ lbs for every inch of width.

By the foregoing statement, it would appear that Biddell's scarifier with points an inch wide will accomplish nearly double the quantity of work, at the same expense of labour, and in the same time as is required for Lord Ducie's cultivator with points two inches wide. There is no doubt but that the land is more thoroughly disturbed with the wider points; but, inasmuch as the operation of perfect scarifying could not, with either implement, be satisfactorily accomplished without a second application, it is contended that it is a more advantageous distribution of the labour to cover a large quantity in a short time, as by the time the second operation is performed, Biddell's scarifier will have thoroughly stirred the whole surface of ten acres in the same time, and at the same expense of horse-power as the Ducie cultivator will have required to accomplish little more than half the quantity.

On the above points of comparison, it appears to me Biddell's scarifier has decidedly the advantage; and to this may be added the facility afforded for adapting the relative depth at which the tines may be required to work, to any inequalities in the general level of the land—an adjustment for which Lord Ducie's implement has no provision. In one point only do I think that the Ducie cultivator has any advantage over the other, and that is, that in the operation of broad sharing near the surface in very foul land, the distance of the tines behind each other admits the rubbish to pass more freely; but for all deep operations, *i.e.*, the usual operations for a scarifier, this advantage is more than counterbalanced by the necessity it involves of constructing the implement of greater length than width, which militates against its even and level working in hard land.

I have been induced to dwell at greater length upon these implements, inasmuch as the importance of the operation of scarifying, as a means of tillage, is increasingly recognised, and the implements for that purpose are at the present time obtaining considerable attention.

An efficient scarifier with sufficient power is calculated, in a great degree, to supersede the frequent use of the plough, and in several instances its operation is attended with better effect.

There can be no question that land is generally in finer tilth after the scarifier than after the plough. How objectionable soever it may be to work the surface of land when it is wet, it is still more objectionable for spring crops, to turn up the wet furrow to receive the seed. It would be frequently impracticable to wait till the season is so far advanced, as for the land to become dry at the ordinary depth of the plough; although, from the effect of frost and exposure to the air, there may often be a fine tilth at the surface, the most advantageous for the reception of the seed: but inasmuch as it is generally the case that heavy land, laid all winter without ploughing; after barley is put in and clover sown, must lie fifteen or eighteen months before it will be stirred again, it is essential that it should be well broken up below at the same time that the fine pulverized soil is retained at the top.

A good scarifier is the only implement that will accomplish this; and it is therefore an indispensable implement on every farm, were it only to enable the agriculturist to get the land into suitable condition to put in the spring crops. Its importance as a cleaning implement has long been acknowledged.

and implements for that purpose are at the present time receiving considerable attention.

An adequate ... sufficient power is available ... great degree ... the implement ... to the ... operation is ... [illegible]

[illegible]

[illegible]

SECTION IV.

THE ROLLER.

The operation of rolling the surface of the ground is for the purpose of breaking down clods, rendering the surface of the land more compact, and bringing it even and level.

This is a necessary practice both upon tillage and grass lands, and is of much utility in both sorts of husbandry. In the former case, it is made use of for breaking down and reducing the cloddy and lumpy parts of the soil, in preparing it for the reception of crops. It is also of great use in many cases of light soils, in rendering the surface more firm, even, and solid, after the seed is put in. Its purpose on grass lands is simply to level the surface.

The first use of the roller must be assigned to a later period of Agricultural History, than most of the implements I have yet described. For its construction was needless in the early days of tillage, when only the richest, most easily tilled lands, were cultivated by the then thinly tenanted earth. The tillage of the heavier soils, the stubborn adhesive clays, would naturally be postponed, until the increasing population rendered necessary their more laborious cultivation. This would, as a natural consequence, suggest the invention of an implement to pulverise, mechanically, the hard clods of such soils ;—and hence arose the construction of the heavy harrows, and, subsequently, the roller; which was made, in the first instance, entirely of wood, then of stone, and lastly of iron.

An implement of so simple a construction, the main purpose of which is to render smooth the surface of arable lands, would not admit, apparently, of much variety in its formation. It is, however, one in which greater diversity of form is found to exist than in many other agricultural machines. Rollers are of various sizes, weights, and lengths; and the material of which they are still made, is occasionally iron, sometimes stone, but most commonly wood. Of these, the first is undoubtedly the best, and particularly for rollers composed of two or more separate cylinders, by which the operation of turning at the ends of the ridges is materially facilitated; and the slading of the earth, which would otherwise take place on the head-lands, not only to their great detriment but to the no small increase of labour to the horses, is thereby prevented.

An ingenious gentleman, the late George Booth, of Allerton, near Liverpool, who to a great love of farming added a tolerable share of mechanical skill, and to both ample means to carry out his various devices, constructed a roller, or rather a series of rollers, on the lever principle, that might be pressed down by weights. He contended for solid rolls of very small diameter, as the most effective in crushing the clods, and throwing the greatest weight on the surface of the ground; from this opinion I venture to differ, believing that whatever advantage may arise from small diameters, it is more than counterbalanced by the difficulty of surmounting clods and other obstacles, and their consequent tendency to drive them before the roller, which would also cause increased labour to the horses. I can only give an idea of Booth's invention from memory; I believe his roller was not more than a foot in diameter. It consisted of five cylinders, arranged somewhat in this manner.

BOOTH'S ROLLERS.

HEAVY ROLLER WITH THREE CYLINDERS.

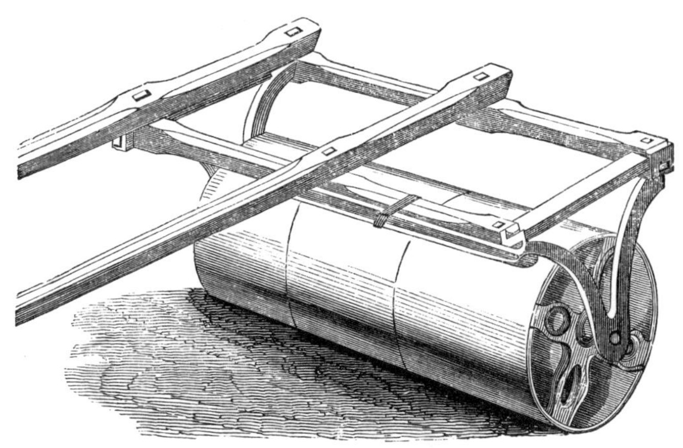

Heavy Rollers.—The heavy roller shown in the above sketch is very effective: it is formed of three separate cylinders, about two feet in diameter, and of the same length; the axis of each being independent of the other. On turning, they consequently revolve in different directions, and thus " slading" at the land's end is avoided.

DOUBLE-JOINTED BARLEY ROLLER.

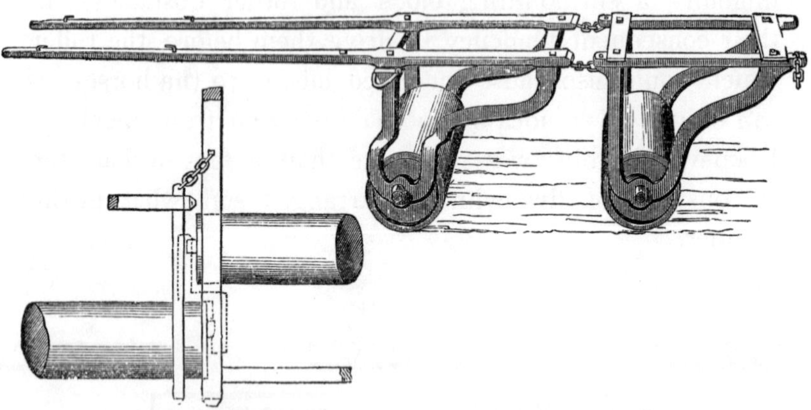

The Double-Jointed Barley Roller is a very useful implement. It is so constructed that the two sides, being separate rolls on distinct frames, may revolve at opposite angles, and, occasionally, or for convenience in travelling, one may be placed behind the other. A plan has for many years been in use in Norfolk of constructing the framing with twisted joints, so that the inner end of one roll shall work behind the end of the other; thus leaving no seam between the rollers. The cylinders of these are sometimes constructed in three parts.

DRILL ROLLER.

Drill Rollers. — These are not of modern invention, having been well known to the farmers of Norfolk and Suffolk, and I believe other counties, for the greater part of a century. The principal improvement they have undergone of late years, has been to render each ring independent of the others, so that the process of turning at the end of the field is facilitated, as in the case of the roller with several cylinders already mentioned. The drill roller is used for the double purpose of crushing clods on rough lands, and making grooves ready to receive the seed of wheat, or other grain sown broad-cast on light soils. It is a capital tool for either purpose. In the first case, it is followed by a harrow, of sufficient weight to lighten up the surface ; in the other, the fine short-toothed harrow, or even a mere bush harrow, will be found sufficient. The less such land is disturbed, after sowing the better ; and the more distinct will be the several rows of corn.

SEAM PRESSER.

Seam or Land Presser.—If a drill-roller be so effective an implement, far more so is the seam or land presser ; inasmuch as its whole force and weight is directed to each furrow, as it is turned over by the common plough. The seam presser is, in fact, an abstract of a drill-roller, consisting of but two cylinders of cast iron, which, following the plough in the furrows, press and roll down the newly turned up earth. It is more particularly useful when applied on ploughed clover stubbles intended for wheat.

CROSSKILL'S PATENT CLOD CRUSHER.

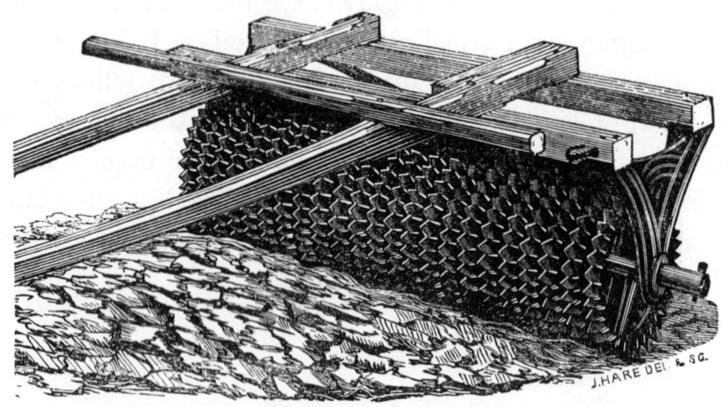

Crosskill's Patent Clod Crusher.—This is, under many circumstances, a valuable implement. It is composed of a series of iron rings, with notched edges set apart from each other about three or four inches. Small cross bars or knives are placed at frequent intervals on the faces of these, and near their outer notched rims, so as to intersect every portion of land over which it passes. Its construction, combined with its great weight, renders it very effective for the purpose which its name denotes. It has been aptly said to be "a roll and a harrow combined." Its use has been found to prevent the ravages of the wire worm; no small recommendation to it.

The inventor of this valuable implement has recently taken out a Patent for further improvements in its construction, the principal features of which consist in an improved form of tooth, for breaking rather than grinding the clods; and in arranging for each cylinder independently to revolve upon its own axis—an advantage which not only increases its efficiency, but materially lessens the power required for its draught.

The roller is an implement which requires more than usual judgment as to the time of its use, and this remark applies with increased force to the one under consideration.

SECTION V.

THE DRILL.

The Drill is an implement for the ready distribution of seed and manure at regular distances. The first, and for ages the only, mode of scattering the seed over the sufficiently pulverized soil was by hand, by which means the seed was of necessity applied with but little precision or economy; and, indeed, in sowing the wet swampy lands under the warm eastern sun, it was hardly possible to distribute the seed in any other way than by the broad-cast system. It was usual, in the East, to prepare the moist soil by treading it with the feet of the ox and the ass. *Isaiah*, xxii. 20; *Matthew*, xiii. 3. A custom still used on low situations in several eastern countries. After the Hindoos have thus deposited their seed, they use a kind of subsoil plough, which passes under and loosens the soil to the depth of about eight inches, three drill's breadth at a time. (*Com. Board of Agr.* vi. 355.)

A rude kind of drill has been used in agriculture from a very remote period. The cultivators of China, Japan, Arabia, and the Carnatic, have drilled and dibbled in their seed from time immemorial. (The Chinese drill, or drill-plough, is noticed in the *Quart. Journ of Agr.* vol. i. p. 675. Gabriel Platte, in 1638--1653, describes a rude dibbling machine formed of iron pins " made to play up and down like Virginal jacks ;" and John Worlidge, in his *Husbandry*, published in 1669, not only advocated the use of the seed drill, but of the manure drill. Evelyn, in the same year, (*Trans. Roy. Soc.* vol. v. p. 1056), mentions with much commendation a drill-plough which had been invented in Germany, whence it had found its way into Spain, and had been noticed by the Earl of Sandwich, the English ambassador, who forwarded it to this country, as the invention of Don Leucatilla. Jethro Tull, at a later period, (1730--1740) devoted all his energies to promote the introduction of this machine, " more especially as it admitted the use of the horse-hoe." The united advantages of these instruments excited in him additional enthusiasm. Hence it appears that the method of sowing corn and other seeds by machines in England is not a modern idea, though the machines have been so much improved within the last century as to make them bear but little resemblance to those formerly in use.

Passing by those of more ancient date, we come to the invention of Jethro Tull, for the purpose of carrying out his system of drill-husbandry, about 1733. His first invention was a drill-plough to sow wheat and turnip seed, in drills three rows at a time. There were two boxes for the seed, and these with the coulters were placed one set behind the other, so that two sorts of seed might be sown at the same time. A harrow to cover in the seed was attached behind.

Jethro Tull also invented a turnip-drill somewhat similar to the other in general arrangement, but of a lighter construction. The feeding spout was so arranged as to carry one half of the seed backwards after the earth had fallen into the channel; a harrow was pinned to the beam, and by this arrangement one half of the seed would spring up sooner than the other, and so part of it escape the turnip fly. When desirable to turn the machine, the harrow had to be lifted and the feeding would stop. The manner of delivering the seeds to the funnels in both the above drills was by notched barrels, and Tull was the *first who used cavities in the surfaces of solid cylinders for the feeding.* Nothing material in the history of the drill occurred afterwards till 1782, and but little progress appears to have been made to that period in drill-husbandry.

About this time Sir John Anstruther, presented the model of an improved drill-plough of his own invention to the Bath and West of England Society, having previously had one in use for eight years without its getting out of order. It was a double drill-plough of simple construction, by which two furrows could be sown at a time, the horse walking between them, and by this means the injury usually done by the horse's feet to the finely prepared mould was avoided. Within the next ten years, twelve patents were taken out for drill-machines, two of which were for depositing manure with the seed; but the most approved appear to have been those invented by James Cooke, a clergyman of Heaton Norris, in Lancashire; and the general principles of these machines, from their simplicity, have been adopted in the construction of some of the most approved of the present day.

The annexed cut of Cooke's drill is copied from the *Letters and Papers of the Bath and West of England Society*, vol. v.

COOKE'S DRILL.

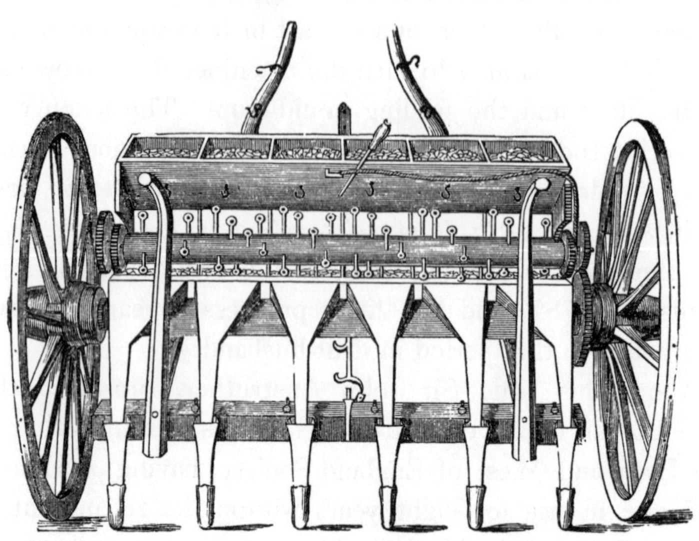

The seed-box is of a peculiar shape, the hinder part extending lower than the fore part. It is divided by partitions, and supported by adjustable bearings so as to preserve a regular delivery of the seed whilst the machine is passing over uneven ground. The feeding cylinder is made to revolve by a toothed-wheel, which is fixed on each end of the main axle, and gears with other toothed wheels on each end of the cylinder; the surface of the cylinder is furnished with a series of cups which revolve with it, and are of various sizes, according to the different seeds intended to be sown. These deposit the seed regularly in funnels, the lower ends of which lead immediately behind the coulters, which are connected by a beam, so as to be kept in an even line, and are capable of being held out of working when desired by a hook and link in the centre. The seed, as it is deposited, is covered in by a harrow fixed behind. The carriage wheels are larger in size than usual, by which means the machine is more easily drawn over uneven ground; and the labour of working is reduced.

About the year 1790, Henry Baldwin of Mendham, a farmer, near Harleston in Norfolk, aided by an ingenious workman named Samuel Wells, improved upon the drill known as Cooke's drill, which by this time was in use in several parts of Norfolk. The improvement consisted—

First—In making a sliding axle-tree, *by which the carriage wheel could be extended at pleasure to the width of the "stetches" or lands, and by which means another box with cups and more coulters could be used. Thus a drill containing fourteen coulters could be enlarged to one of eighteen or twenty.*

Second—In making self-regulating levers, *to which the coulters were attached; this was done by hanging each coulter on a distinct lever, placed at right angles with the cross-bar of the framing, upon which each lever was made to swing by an ordinary hinge joint, and had a moveable weight at its opposite end, to press the coulter into the soil.*

By the levers being thus contrived to work independently of each other, they accommodated themselves to the irregularity of the surface of the land, and the impediments they might meet with, without disturbing the whole. The above were two very important improvements; and they are both in use to this day.

SMYTH'S SUFFOLK CORN AND MANURE DRILL.

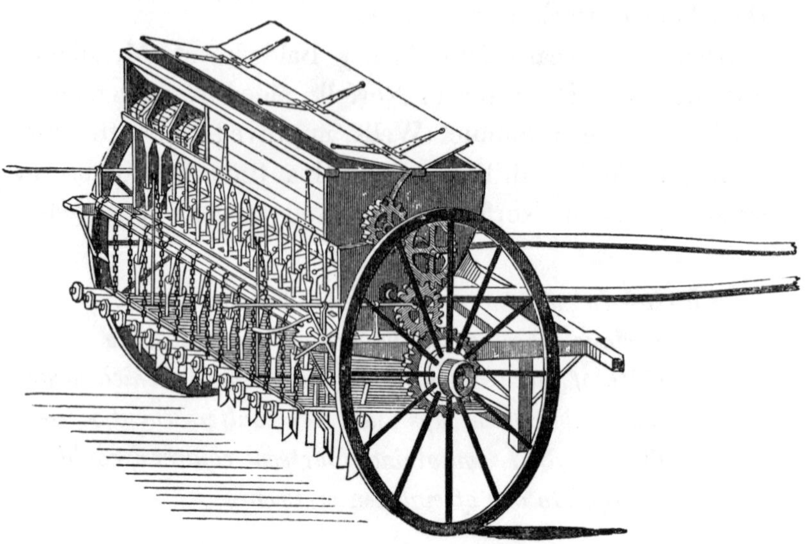

Suffolk Corn and Manure Drill.—Following the improvements just referred to, are those by James Smyth of Peasenhall, and his brother Jonathan Smyth of Swefling, who have been engaged in the manufacture upwards of forty years, and by whose unremitting attention to the practical operations of this valuable machine it has been brought to a high degree of perfection. A brief summary of their improvements is as follows :—

First—A mode of adjusting the coulters to distances apart from each other, from four and a half inches and upwards.

Second—An improved manure box, and cups for the delivery of manure with the corn.

Third—A plan to drill in manure and corn, and sow small seeds at the same time.

Fourth—The swing steerage, by which means the man attending the drill can move the coulters to the right or to the left hand, so as to keep the straight and parallel lines for sowing the seeds.

Fifth—Various improvements in gearing and driving the wheels, barrels, &c.

GARRETT'S PATENT DRILL FOR GENERAL PURPOSES.

SECTION OF GARRETT'S PATENT DRILL.

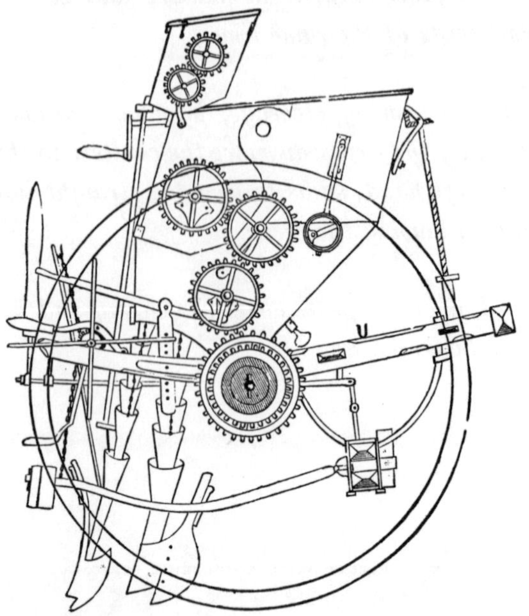

These machines continue to be made by the Smyths of Peasenhall and Swefling, and by Garrett and Son, of Leiston, in Suffolk, which latter have also added some improvements to admit of the adaptation of the same drill to various purposes, for which they have obtained a patent. These improvements consist—

First—In the application of a double-actioned stirrer to the manure-box, which, having a perpendicular and also a revolving movement, constantly disturbs the manure, and presses it forward into the departments for the depositing barrel to act upon it, and by which means the manure, even when coarse, or badly prepared, is equally distributed.

Second—A contrivance, by which the difficulty arising from the different weight of seeds when two or more descrip-

tions of seed are to be sown together, and which has a tendency to cause them to be distributed in unequal quantities, is avoided. This is accomplished by an arrangement which allows the small heavy seeds, as clover and trefoil, to be delivered from cups, while the lighter seeds are, by the same operation, brushed out of a separate compartment in the box down the same conductors as the other seed.

The Suffolk drill is now the kind in the most general use throughout the kingdom, and is adapted for drilling corn either on level lands or on ridges, and on all descriptions of soil. It is, as stated in the previous description, furnished with *independent levers*, by which the coulters are *each* readily and separately made to avoid any rocks or irregularities of the ground, and a press-bar, extending over the entire width of the machine, to force the coulters, in case of need, into hard ground, with a varying degree of pressure, according to the texture of the soil.

The coulters can be set so as to drill the corn at any width, from four inches to a greater distance; they also, if required, readily allow of the introduction of the horse-hoe; and, from being placed in double rows, they admit, when at work, large stones to pass between them of a size that, under the old plan of placing the coulters in one line, would break or stop the machine. The most complete drills are furnished with the "*swing steerage*" before referred to, by which the drill-man keeps the rows at exact or even distances from those which have been previously drilled. The "corn barrel" is made to deliver from two pecks to six or seven bushels, or strikes of seed, per acre; and they are furnished with an additional barrel for drilling turnips and mangel-wurzel. These barrels, by a simple yet efficient "regulator," are kept on unequal hilly ground at the same level; so that the grain is evenly delivered, in whatever situation the drill may be placed.

A "*seed engine*" is also sometimes added to the common corn drill, by which the grass seeds and clover are sown at the same time as the corn, and each kind of seed, if required, separately. By this plan any quantity per acre of the seeds may be much more evenly distributed than by mixing them up together. For these seeds, being of different sizes and weights, are, in the ordinary seed engines, very apt to separate in the boxes; and thus the brushes too often deliver them in unequal proportions.

The weight of these drills necessarily varies with the number of coulters; ranging from three to ten cwts.; they are drawn, according to circumstances, by one, two, or three horses; the sliding axletree, allowing the addition of any number of coulters, adapts the drill to different breadths of land.

The manure-box may be taken on or off at pleasure. It is a simple yet accurately-working apparatus for delivering the manure, which it does with great evenness, and in quantites varying as the "slip" is placed, from six to eight bushels per acre. In the best drills, also, a very important improvement has been made within the last few years, which consists in the use of separate coulters for manure and seed. The manure is now deposited according to the mode preferred by the cultivator, not only from two to three inches deeper in the ground than the seed, but from ten to twelve inches in advance of it, so as to give the soil time to cover the manure, before the next coulters deposit the seed;—whereas, on the old plan of depositing the seed and the fertilizer together down one pipe, an evil was liable to arise; when it was used with some of the more powerful artificial manures, the seed and the manure were too close together, and the manure was not dropped with certainty in its best position, *under* the seed.

BEDFORDSHIRE DRILL.

The next machine which claims notice is the *Bedfordshire drill.* This drill was the invention of Robert Salmon, of Woburn, who obtained for it the premium given by the late Duke of Bedford, at his annual sheep-shearing about thirty years ago.

It is an ingenious and a simple machine, and so contrived that the drill-man can easily direct its course while he is attending to the cups, and otherwise superintending its operation. In the arrangement for its guidance consists its principal advantage. It was improved by two brothers, James and Thomas Bachelor, farmers and mechanists, at Lidlington, near Bedford; and afterwards by the present makers, Samuel Hensman, of Ampthill, William Hensman, of Woburn, and William Smith, of Kempston, with some others in the county of Bedford. The form of this drill will be seen by the engraving.

The seed-box is suspended upon two centres, one at each side; on these it swings so as to keep its level position as the drill moves up and down hill, or over ridges. Sometimes, instead of the box being made to swing upon centres, it is fixed so as to be altered, as occasion may require, by an adjusting screw and crank placed in a convenient position for the drill-man to regulate. The seed corn is taken up by iron cups fixed on circular plates and delivered into funnels, from whence it descends to the ground; these plates are centred upon a spindle, which revolves by being connected with the nave of one of the carriage wheels. The coulters are forced into the land by an equal pressure upon each from the centre of the carriage, on which nearly the whole weight of the drill rests. The steerage is effected by a pair of light wooden bars, attached to the axle on which the carriage wheels run. These have a cross bar at their ends, to which small handles are attached, so that the man may guide the drill to a nicety, whether he be at the right or left side of it. Drills on this principle of steerage are made with four, six, or eight coulters; the two larger sizes are in general use in Bedfordshire: those with six coulters are considered the best for heavy or hilly land; those with eight coulters for light and level lands.

The Bedfordshire drill, which has been thus described, is an exceedingly useful little machine for sowing corn; but it is not suited for the combined purpose of sowing manure and corn at the same time, as the weight of the additional box, when first charged with manure, would be so great as to press the coulters much deeper than is advantageous for depositing the seed, and when nearly emptied, the converse would probably be the case. The total weight of the machine, with the additional loaded manure-box, would also render it unwieldy for a man to turn at the ends of the land.

Robert Maynard, of Whittlesford, Cambridgeshire, an ingenious mechanist, to whom the agriculturists are indebted for several improvements in drills and other machinery, has, by the simple contrivance of placing a roller immediately in advance of each coulter, remedied this defect, as these rollers not only prepare the surface of the soil for the most favourable operation of the coulters, but at the same time secures their penetrating the ground to a uniform depth.

In his drill the seed-box is a small compartment in the manure-box; its total weight, when loaded with about five bushels of manure, the quantity it is calculated to hold, is not much more than that of an ordinary corn-box, and by placing it somewhat more in advance than the common seed-box, the operator is relieved from any undue weight, which is thus thrown upon the wheels.

Lord Western's Patent Drill.—The improvements to the drill invented by this nobleman, consist—

First—In the application of improved metallic hinge-joints to the moving levers, so as to insure their continuing parallel with each other, and in a line with the course of the machine.

Second—By the addition of a pair of thin-edged wheels and a fore axletree to the machine, by which it can be guided with certainty in any direction.

Third—A peculiar apparatus for steering the machine,

consisting of a shaft passing through a socket, fixed in a plummer-block to the foot-board; its outer end being attached to the fore axle by a universal joint, the other end having a wheel with handles fixed thereon, by which the machine is readily steered in its course.

Fourth—The application of improved iron sockets to the sides of the levers, as a means of fastening the coulters, instead of weakening the levers by passing through them as heretofore.

LORD WESTERN'S PATENT DRILL.

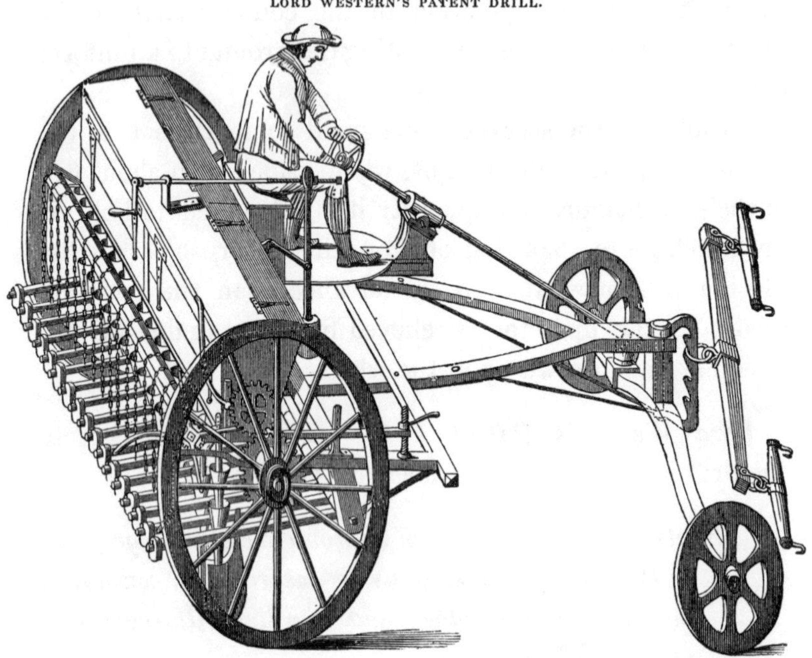

Lord Western's drill also admits of horse hoes being substituted for the drill coulters, and he has had the hoeing between young plants exceedingly well performed in this manner on his own farm, to which he invited public inspection.

GROUNSELL'S PATENT DROP DRILL.

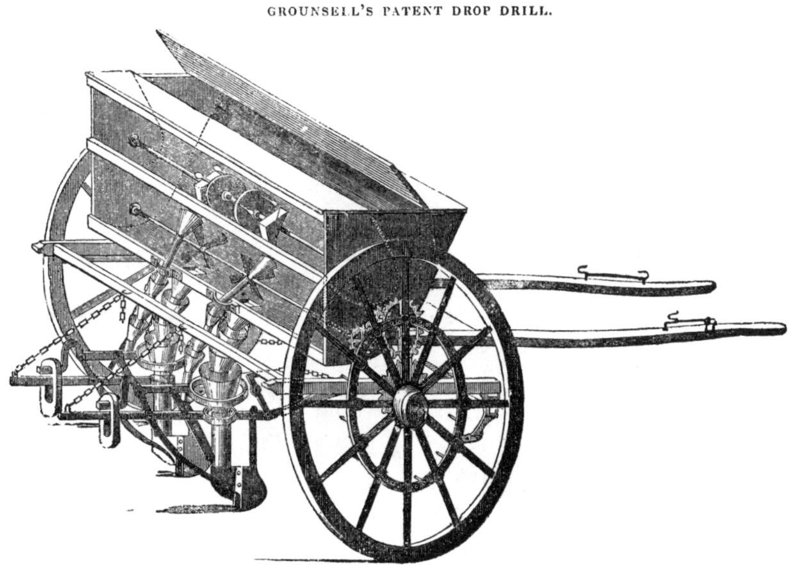

Grounsell's Patent Drop Drill—This drill is for the purpose of depositing corn, grain, pulse, and manure *at intervals*, the distances of which may be regulated at pleasure.

To effect the purposes above mentioned, a circular iron ring is fixed about midway between the nave and rim of the drill carriage-wheel. In this there is a number of holes to carry a series of studs, which may be varied according to circumstances; and, as these studs come in succession, when the wheel turns they open valves for the delivery of the corn and manure, which close again immediately the stud has passed. A further improvement consists in the adoption of projecting arms or shovels, to draw the manure and corn to the funnels, instead of taking the same up in cups, in the way adopted in other drills.

HORNSBY'S PATENT DROP DRILL.

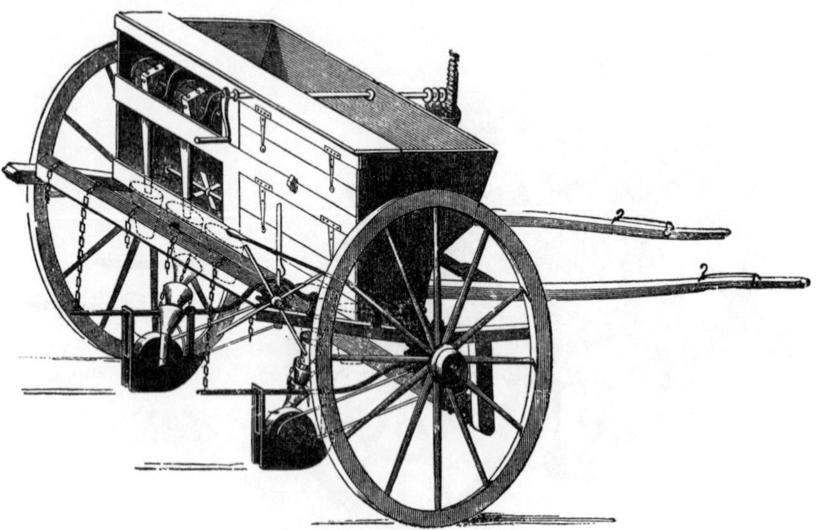

The next drill is *Hornsby's Patent Drop Drill.* This drill is also intended for dropping seed with manure at intervals, but the construction of it is very different to the one last mentioned. In this, the manner of regulating the delivery is by having a coulter of a peculiar form inside, in which a circular box revolves on an axle which passes through one side. This box is divided into compartments closed by small doors, which are kept shut by a spring to each; the compartments in the box are supplied through a series of funnels, the end of the lower one entering one side of the box below the centre.

On the machine being moved forwards, this box revolves by means of appropriate cog-wheels; and as each spring arrives at the ground, the door to which it is attached opens, and the contents of that compartment are deposited, to be again replaced when it arrives at the part of its rotation at the end of the funnel, and so on successively.

Turnip and Manure Drill for Ridge Work—This drill, as constructed by Garrett and Son, who have furnished me with the following description, has the same advantages in the manure department as their drill for general purposes, before described, with this additional one, viz., that the regulation of the quantities of manure to be deposited is placed immediately under the management of the attendant, who, as he follows the drill, may so adjust its distribution as to admit larger or smaller quantities, as may be required on varieties in the soil. In order to accommodate the coulters to irregularly ploughed ridges, a pair of rollers are attached to a fore carriage, so as to form and press the land properly for the deposit of the manure and seed, and that which has hitherto been found a difficulty, viz., the keeping the coulters on the tops of the ridges, is here remedied by each lever being easily altered, by a parallel motion, to move wider or narrower, independently of each other, so that the seed coulter advances in the exact line of the manure coulter. Rakes are attached, which add greatly to the effective working of the drill, the foremost filling up the channel left by the large manure coulters, thus permitting the seed to be deposited in fresh-stirred mould directly above the manure, while the hinder rakes serve to cover the seed.

Having described the larger drills, I pass over the numerous varieties of small drills, as they are simply modifications of the larger ones ; and the descriptions already given, embrace most, if not all, the mechanical principles and contrivances applied to these machines.

From this brief enumeration the farmer will see that the

modern drill makers have not neglected their duty, in the adoption of many improvements calculated to simplify and render more serviceable the common and the manure drill; and I am gratified to be able to add that there is now every prospect of their skill and enterprise being rewarded, for I find, from an eminent maker, that the demand for manure drills has within the last two years been greater than was ever remembered.

The chief advantages of the use of the drill, "are the regular deposition of the seed at an uniform regulated depth, from which arises a considerable saving of seed; and the facility afforded for cleaning the land either by the hand or horse-hoe." The importance of these results is rapidly becoming generally understood; and the results of extensive experiments, have proved that by the use of this machine, combined with careful hoeing and weeding the crops, a saving even of half the usual quantity of seed now used by the drill may be effected. The use of the manure drill, and of those fertilizers prepared expressly for its use, cannot be too strongly urged upon the farmer of the upland soils, far away from supplies of manure: Cuthbert Johnson, in his *Farmers' Encyclopædia* judiciously observes, "remember, that it is not only the first cost of all manures which makes them expensive, but the comparative *labour* saved in their application, which must also be taken into the account, when the cultivator is estimating their value. And further, let him remember that the best and richest farm compost is likely to convey to his fields a multitude of seeds, the cost of whose removal too rarely forms a portion of such comparative estimates."

DIBBLING MACHINES.

THE subject of drilling by machinery naturally suggests the consideration of whether the operation of dibbling may not be similarly accomplished. Many ingenious contrivances have from time to time been projected for this purpose, and several patents have also been obtained, but I am not aware of any that have long been successfully and advantageously used; it is, therefore, unnecessary to describe their peculiarities.

We have noticed two at the different exhibitions of the Royal Agricultural Societies of England. One, the invention of Thomas Wedlake, of Hornchurch, and the other by W. Lewis Rham, Rector of Winkfield; but, at present, I have had no opportunity of ascertaining whether in practice they justify the good opinion formed of them; knowing, however, the ingenuity which has been brought to bear by both these gentlemen upon other articles of agricultural machinery, I have no doubt but it will be exerted with success in bringing these implements to perfection.

The great practical difficulty attending the construction of these machines, rests on the circumstance, that, in order to ensure accuracy in the deposition of so small quantities of seed, the mechanical arrangements require to be delicate and minute; these having to be placed so near the soil, in fact, almost to work in it, the dust on light soils, and the adhesive character of the heavier soils, are alike destructive to the maintenance of the more delicate parts in their proper working positions. The motion required for the operations of forming the hole, by giving the dibble a twist or half turn before leaving it, and the necessity to deposit the seed precisely in the spot prepared for its reception, necessarily involve a somewhat complex mechanical arrangement; in the machine alluded to of W. L. Rham, these difficulties are more nearly overcome, than in any other I have before seen.

SECTION VI.

THE HORSE HOE.

For this valuable implement of agriculture the farmer is indebted to the justly celebrated Jethro Tull. Previous to his time, we search in vain in the works of agricultural authors for the slightest allusion to such an instrument. The production of the horse-hoe, indeed, seems to have been almost a natural consequence of the adoption of the drill system, for which also the cultivator is mainly indebted to Tull. He gave in his *Husbandry*, more than a century since, an engraving of a horse-hoe of his own invention, which resembles a common, rudely-shaped swing plough, with the mould board omitted, and the share having a cutting edge turned up on its landside. A variety of improvements, were gradually made in the construction of this implement: I proceed to notice those which are now considered to be the best.

The advantages which horse-hoes possess over the hand-hoe are very fairly stated by the late Francis Blakie: he remarks, "In many cases the hand-hoe may be used to advantage, and should then be so used. But generally speaking, the hand is not so efficient as the horse-hoe. Expedition is a most material point in all processes of husbandry, carried on in a variable and uncertain climate, and it frequently happens, that hoeing, in any way, can only be executed to advantage, in a very few days in spring: hence the horse-hoe has a most decided advantage over the hand-hoe, for a man will only hoe about half an acre a day with the latter, while, with the former, a man and a boy, with one horse, will hoe eight or ten acres a day, and that in a more effectual manner." (*On Farm-yard Manure*, p. 39.)

The following is an engraving of the implement invented by him, which bears his name.

BLAKIE'S INVERTED HORSE-HOE.

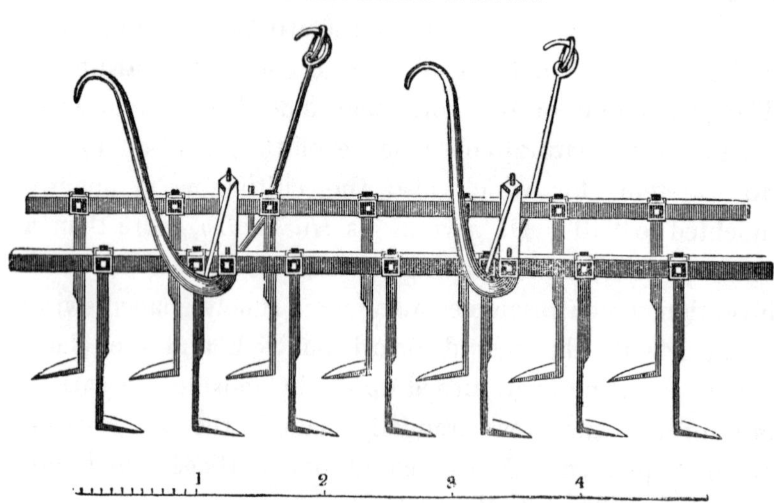

Blakie's Inverted Horse-hoe.—This excellent implement, which has the merit of being the first successfully employed to hoe between several rows of turnips at once, was introduced by Blakie, when manager of the estate of the late Earl Leicester, and obtains very general use in some parts of Norfolk. It consists of two beams or bars of iron, one placed in advance of the other, upon which the hoes are placed alternately; the hoes instead of being formed by a triangular blade attached to a stalk springing from the centre (see figs. *a b*, p. 123.) are small blades of steel attached to a stalk, and inverted, or thrown backward from it, in the manner shown by the engraving; thus, instead of the stalk passing along the middle of the space between the plants, which would have a tendency to drive the earth from the centre to the right and left, the stalks of each pair of hoes pass closely to the plants, and the earth cut by them passes between them. Blakie, in his own description of it, observes, the idea first struck him on observing a large proportion of the plants buried by the operation of the hoes formerly in use. The frame has two handles, and the fore-part of it is attached, by draft irons, to the axletree of a drill, or any light carriage. This implement is well adapted for cleaning between rows of plants growing at narrow intervals, within which it may be worked in perfect safety when in their infant state.

GRANT'S HORSE-HOE AND MOULDING PLOUGH.

Garrett's Horse-hoe and Moulding Plough.—The form of this implement will be seen by the engraving; it is made entirely of wrought iron; mortices are made through the beam to receive two iron bars, upon which the frames supporting the hoes may be adjusted to cut any given width, or at any desired interval. By substituting mould boards in the place of the wrought-iron frame for the hoes, this implement is converted into a moulding plough. The inventor has since obtained a patent for *an Improved Lever Horse-hoe*; this consists of a number of hoes so arranged on a frame as that, by means of a compound lever, similar to the one described and shown at p. 129, on his patent horse-rake, all the hoes may be instantly raised should they become clogged with the weeds or rubbish. Each hoe is placed on a separate lever, as the tines of the patent rake.

MORTON'S, OR LORD DUCIE'S HORSE-HOE.

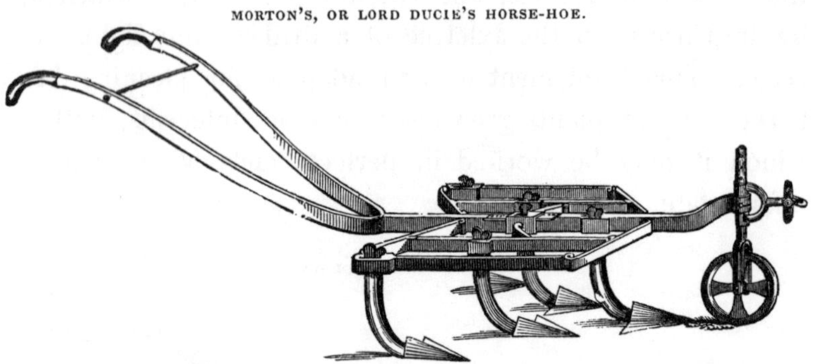

Lord Ducie's Parallel Expanding Horse-hoe.—The description of this implement, by the inventor, states that its general uses are for hoeing drill crops but that it can be converted into a light scarifier, by taking out the hoes and putting in tines, sent with the machine. It is constructed principally of wrought iron, but the cutting edges of the hoes are steel and hardened so that they will be always sharp: it has five tines, and can be regulated to any width, from twelve to twenty-seven inches, with the greatest facility, so that the hoe shall always present its edge in a straight line to what it has to cut: this is effected by the support of each hoe moving parallel with the beam: it is worked on the principle of the parallel rule; the machine has one wheel in front, with a tiller for the horse to yoke to: the depth it enters into the ground is regulated by raising or lowering the wheel; there is a pair of handles for the man who attends the machine to steady it by.

CLARKE'S UNIVERSAL RIDGE PLOUGH.
(*Shown as a Horse-hoe.*)

Clarke's Universal Ridge Horse-hoe.—This is a very ingenious contrivance of John Clarke, of Long Sutton, for carrying out the several operations of ridge culture, and for which the English Agricultural Society awarded its silver medal at its first meeting.

It is adapted for the uses of a double tom, a moulding plough, a broad share or cleaning plough, and a horse-hoe. It is only as fitted for the latter purpose that we have now to describe it, having previously referred to it under the article PLOUGH. (*Vide* p. 46.)

To the frame of this plough is attached a pointed share which serves as a hoe for the centre of the furrow: a moveable jointed frame is affixed to the beam, which is readily adjusted to any given width: to this is fastened, when it is intended to hoe plants upon the ridge, the stalks of two curved hoes, as shown in the drawing; when used upon flat work, the flat hoe *a*, should be substituted for the curved or inverted hoe *b*. This forms a very perfect and simple horse-hoe.

WHITE'S DOUBLE ACTION TURNIP-HOE.

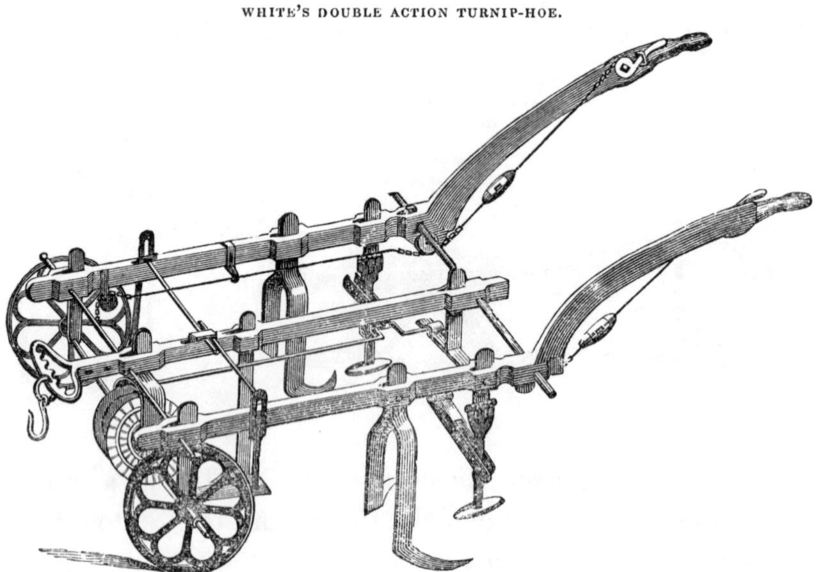

White's Double-action Turnip-Hoe.—This implement is intended to hoe turnips sown either broadcast, by the drill, or on the ridge. In addition to the hoes intended to pass in the direction of the ridge, or furrow, or between the rows of the plants; it is furnished with an apparatus which, by means of a crank put in motion by the fore-wheels, two hoes are made to traverse the rows, and thus to cut out the plants when it is desired to thin them on the row. It admits of adjustment to suit rows of any distance, from fifteen to thirty inches wide, and the cross-hoes may be made to operate so as to leave the plants on the rows at distances of either nine, twelve, or fifteen inches apart; it may be used with one man and one horse, and by the substitution of spear-footed tines in the place of the hoes, it becomes an efficient light scarifier.

Huckvale's Patent Turnip-hoe.—This implement is intended to accomplish a similar operation to that described in White's implement; but the hoes for cutting out the turnips on the rows are, by a simple arrangement of two cogged-wheels, made to revolve in the direction across the ridge, while the body of the implement, containing the hoes for cleaning between the rows, is drawn forward by a horse. The writer has not seen either of these implements in operation but has little doubt they would be found effective to accomplish the object for which they are designed.

The great advantage alluded to by Blakie, of rapid execution in the hoeing of crops in the spring, is hardly, even at this time, sufficiently appreciated. A perfect implement to hoe several rows at once, has been long a desideratum, and the want of it is probably the cause of the system not having been generally adopted. Blakie's, however good, when used on light land of a generally even texture, was not suited alike for light and strong soils, as it admitted of no other adjustment to force the hoes to take hold of hard land, than the pressure of the handles. Garrett's patent hoe, which I have now to describe, and Grant's, before alluded to, supply the remedy to these defects, and either of these implements may be considered perfect.

GARRETT'S PATENT HORSE-HOE.

Garrett's Patent Horse-hoe.—This horse-hoe, invented by Garrett and Son, of Leiston, is suited to all methods of drill cultivation, whether broad, stetch, or ridge ploughing; and is adapted to hoeing corn of all kinds as well as roots. The peculiar advantages of this implement are that the width of the hoes may be increased or diminished to suit all lands, or methods of planting; the axletree being moveable at both ends, either wheel may be expanded or contracted, so as always to be kept between the rows of plants.

The shafts are readily altered, and attached to any part of the frame, so that the horses may either walk in the furrow, or in any direction, to avoid injury to the crop.

Each hoe, or each pair of hoes, works on a lever independent of the others; so that no part of the surface to be cut, however uneven, can escape; and in order to accommodate this implement to the consolidated earth of the wheat crop, and also the more loosened top of spring corn, roots, &c., the hoes are pressed in by different weights being hung upon the ends of each lever, and adjusted by keys or chains, to prevent their going beyond the proper depth.

That which has hitherto been an objection to the general use of the horse-hoe in this is avoided by adopting a mode of readily shifting the hoes, on a plan similar to that of the steerage adopted in drills, and before described at p. 105, so that the hoes may be guided to the greatest nicety. This implement is so constructed that the hoes may be set to a width, varying from seven inches to any wider space; the inverted hoes are preferred when the distance between the rows is sufficient to admit a pair of them; otherwise, triangular or arrow-shaped hoes may be substituted, or any other form that may be considered best for the purpose.

Two points in this hoe are worthy particular notice: the one being that the blades of the hoes are made entirely of steel, and are attached to the stalks so readily, that as they may become damaged or worn out, they may be replaced by the operator without difficulty; the other, that the position of the frame admits of easy adjustment, so that, according to the texture of the soil, the cutting edges of the hoes may assume a position more or less inclining to the work.

SECTION VII.

THE RAKE.

An instrument similar in some degree to the common English rake has been in use from a very early period by the cultivators of most parts of the earth. The invention of rakes worked by horses, is, however, an improvement of a modern age; and it is to this last named variety that I shall chiefly refer in this Section.

IMPROVED DRAG RAKE.

The Drag Rake, in its simplest form, is merely a long cross-head, with a row of teeth placed in it. In some these are straight; but those most in use are bent, with their points projecting forward. A very excellent and light hand implement, having the teeth of steel, and made with screws, so as to admit of their being easily replaced in case of accident, is well known as Badgley's improved drag rake. And so light to handle, and convenient is this, that girls of fourteen and sixteen years of age can readily use it.

SUFFOLK HORSE DRAG RAKE.

Other rakes similar in kind but of a larger description are made, having two small wooden wheels attached to the cross head, which render them manageable by women or boys. Further addition having been made to them, they are now sufficiently strong and answer better to be worked by a horse. They are used on fallows to remove the couch grass, and get together the rubbish, or, in harvest, time to collect the loose corn which may have escaped from the scythe or sickle. In order to clear them readily, there are different contrivances. One of the most simple is exhibited in the above sketch, where it will be perceived that by lifting the handle the teeth are brought between two iron bars which constitute a part of the framing; by which means all the rubbish is stripped off from the teeth of the rake.

This simple form of implement is in very general use in Suffolk and other eastern counties, but from the circumstance of its tines being rigidly fixed in a cross head, they are continually liable to be broken in uneven or stony land.

WEDLAKE'S HORSE HAY-RAKE.

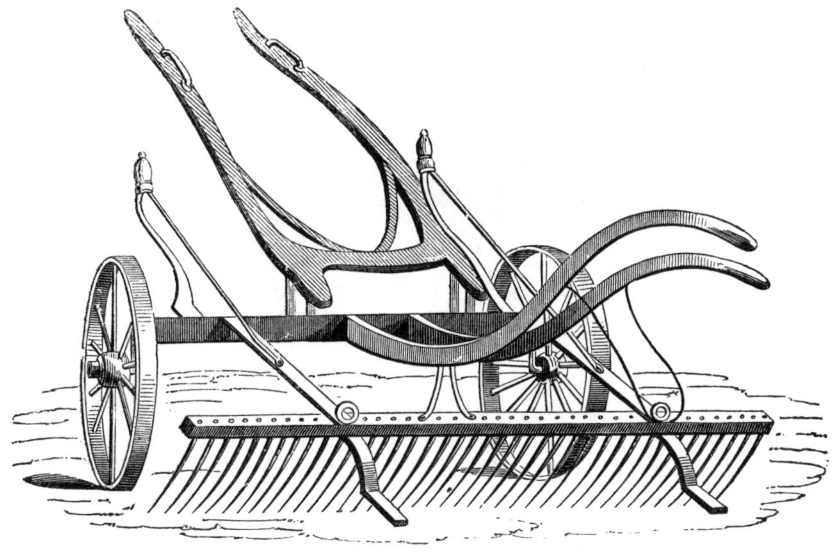

The above is a sketch of "*Wedlake's Horse Hay-Rake.*" The weight of this rake is balanced upon the carriage by two heavy balls projecting in front of it; so that a slight lifting power applied to the handle will raise it from the ground, and disencumber it of the hay or stubble it may have gathered, while, at the same time, the cross-head and teeth being nearly equally balanced with the balls in front, prevents the instrument from sustaining any damage when the tines may come into contact with stones or other inequalities on the surface of the land over which it has to pass.

This rake obtained the commendation of the Committee on Implements at the meeting of the Royal Agricultural Society of England at Cambridge.

EAST LOTHIAN STUBBLE RAKE.

The East-Lothian Stubble-rake, is a machine not so well known in this country as its merits deserve. Its advantages over those previously described are as follow:—It has each tooth placed in a separate head, which, working upon a centre like the levers of a drill, adapts itself to any inequality in the ground. A bar the length of the harrow is firmly fastened to the handles, and from this bar each lever is suspended by a few links of chain. When it is necessary to clear the rake, the handles, on being elevated, lift all the levers between a framing of light iron rods. The above is a sketch of this implement as generally used.

An ingenious practical farmer, John Sayer, of Bodham, in Norfolk, made considerable improvements upon this rake, by altering the form of the teeth to avoid tearing the land; and, in order to effect more work, without increasing the width of the rake, the naves of his wheels were made to project inwards, so that two additional levers could be introduced working quite close to the spokes.

GRANT'S PATENT LEVER HORSE-RAKE.

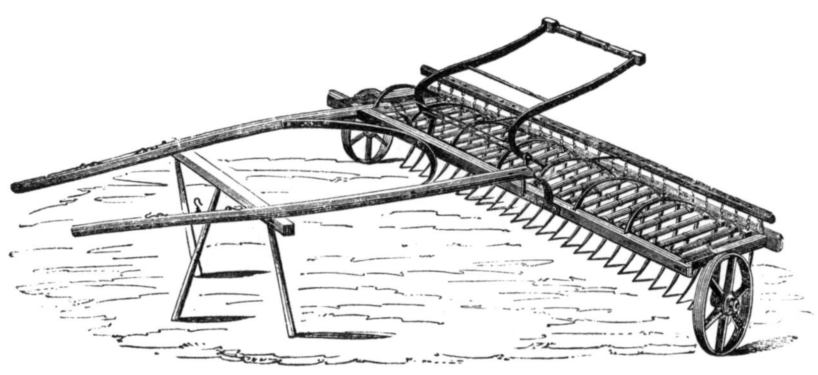

Within the last few months a very improved implement of this character has been invented and patented by J. C. Grant, of Stamford, and it obtained the prize of the Royal Agricultural Society of England, at its meeting at Liverpool, where it attracted great attention. Its advantages consist in the adaptation of a compound lever, by which the whole row of tines may be instantly raised, and as quickly resume their proper position. The form of the teeth being such as to describe part of a circle, the centre of which is the axis of the separate levers to which they are attached, each portion of the curve is successively brought into a vertical position, and thus the teeth are rapidly disengaged from the material collected, so that, without stopping the horse, the process of collecting is resumed, leaving no interval beyond what is requisite for the deposit of the hay, corn, or stubble previously collected.

Several minor improvements are included in the patent,

but as these mainly refer to modes of construction, it is not necessary here to particularise them.

The foregoing sketch will show the general construction of the improved implement.

SALMON'S HAY-MAKING-MACHINE.

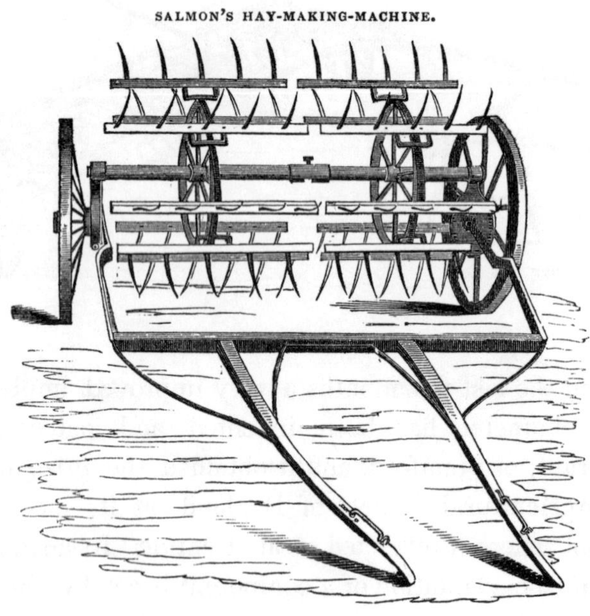

In connection with the horse rake may be mentioned the *Hay-making machine* invented by Robert Salmon, of Woburn, and patented in 1816. This, as will be seen by the figure, is a series of rakes revolving upon two skeleton frames, to which motion is communicated by cog wheels attached to the naves of the wheels on which it travels. It has undergone considerable improvements by Thomas Wedlake, of Hornchurch, from one of whose implements the drawing of the above is taken. His improvements consist in forming the cylinder in two parts, each of which has motion independent of the other, and in placing the tines or rake teeth upon a bar, which being supported by a spring, will yield to the obstructions caused by any sudden unevenness of the surface of the ground, and will return again instantly to its original position. Its object is to spread the hay, and by thoroughly separating its parts, continually to expose them to the sun and wind, which it does so perfectly that the hay is fit to cart much earlier than by the common process of shaking it by hand. To the practical agriculturist, it will not be necessary to remark on the advantages accruing from the means of hastening, if only by a few hours, the process of hay-making; but I may add that it is universally admitted that this is a machine calculated to secure the greatest economy both of time and weather, and therefore one of great value to the agriculturist.

SECTION VIII.

THE THRASHING-MACHINE.

Having thus far described the machines principally in use for the tillage of the land, and for gathering and harvesting the crop, I now arrive at the second part of my subject, the description of those machines which are employed for preparing the stored crops for use.

Foremost amongst these stands the thrashing machine; not only from the circumstance of its evidencing a higher character of mechanical arrangement than at present attaches to any other implement of agriculture, but from its importance as a means of economising both labour and time in an operation which, as performed by the more common instrument, the flail, is tedious and imperfect.

In reverting to the practice of ancient times, to accomplish the separation of the corn from the straw, it is scarcely necessary to state that of all the modes then adopted, "the bruising with the cart wheel," "the sharp thrashing instrument having teeth," "the trampling under the feet of the unmuzzled ox," or "the rollers plain or fluted" mentioned in the practice of continental and eastern agriculture, the flail alone remains in use; and it is with this instrument, preserving very nearly its original form, and by which, till very lately, the entire growth of corn and seeds in this kingdom was thrashed, that a comparison of the more modern inventions of the present and last century will have interest.

That the flail may be made thoroughly to effect, though at great cost of labour and time, the purpose of clearing the grain without damage either to the corn or the straw, is a point no one will be inclined to dispute. But the disadvantages attending its use are not confined to waste of labour and time; for though it may be granted that the operation if properly performed may be perfect, it is difficult to secure its proper performance; and the agriculturist, whose journeys to his barn have from time to time interrupted his surveys of other not less important agricultural operations, can fully testify to this. It is evident that the latter part of the operation of thrashing by the flail will require much more labour to produce a given quantity than its earlier stages; and hence the straw is frequently passed away partially thrashed, in order to procure a greater bulk in a given time. Nor are these disadvantages all that attend the flail. Constant inspection may, perhaps, to some extent, remedy them; but no attention will altogether suffice to remove the temptation to dishonesty which is continually presented where large quantities of grain are under the eye and in the power of those to whom a small portion is of great importance; and hence arises, even when undetected, and often, indeed, when not committed, a cause of great temptation on the one hand, and injurious suspicion on the other.

To overcome these evils, prejudicial not only to the true economy of the farm, but to those feelings of confidence which, justly to sustain the social bond, should ever exist in the relation between the labourer and his employer, the attention of our enterprising neighbours in Scotland was first successfully directed to the construction of machinery; and in 1732, Michael Menzies, a gentleman of East Lothian, invented and patented a machine for thrashing grain, about which period Jethro Tull had in England made similar, although less successful, attempts. I regret that, as nothing but the bare record of Menzies' invention is enrolled in the Patent Office, not having been able to learn more of this, the germ of thrashing-machine-invention, than that it was a contrivance by which a series of flails were made to revolve upon a cylinder; but I am pleased to be able, in some degree, to redeem it from the "condemnation of faint praise" with which we find its memory *generally* accompanied, by reference to the report of a committee appointed by the Society of Improvers in Scotland, to inspect its operation and determine upon its merits. This committee, after various trials, reported it to be "their opinion, that the machine would be of great use to farmers, both in thrashing the grain clean from the straw and in saving a great deal of labour; for one man would be sufficient to manage a machine which would do the work of six." They further recommended the Society "to give all the encouragement they could to so beneficial an invention, which, being simple and plain in the machinery, might be of universal advantage." The Society approved of the report, and acted upon the recommendation.

During the next period of twenty years there appears to be scarcely any other attempts to carry out the object of thrashing by machinery; but, in 1753, Michael Sterling, a farmer in the parish of Dumblane, Perth, applied the principle of the mill in common use for hulling flax, to this purpose. This mill had a vertical shaft, with four cross arms inclosed in a cylindrical case, three feet six inches high, and about eight feet in diameter. The shaft was made to turn with considerable velocity, and the sheaves were gradually let down from an opening at the top; the grain and straw, after being subjected to this beating, were then pressed through an opening in the floor, where rakes and fanners completed the separation of the grain from the straw and the chaff. It was, however, found that this machine "broke off the ears of barley and wheat instead of clearing them of the grain, and that, at best, it was only fit for oats."

It is curious to trace the various plans by which the desideratum of thrashing by machinery was attempted, and a slight sketch of a few of these will suffice satisfactorily to show that in following out the principles which distinguish Meikle's machine (hereafter to be mentioned), little of value has been lost to the public in those which have fallen into desuetude. In 1772, Alderton and Smart, two gentlemen, residing in Northumberland, invented a machine, by which the sheaves were carried round between an indented drum of six feet diameter, and a number of fluted rollers, which, pressing by means of springs against the fluted concave, rubbed out the corn from the ears; and, in 1785, William Winlaw, of Mary-le-bone, patented an invention which he denominated a "mill for separating grain from straw."

WINLAW'S MILL.

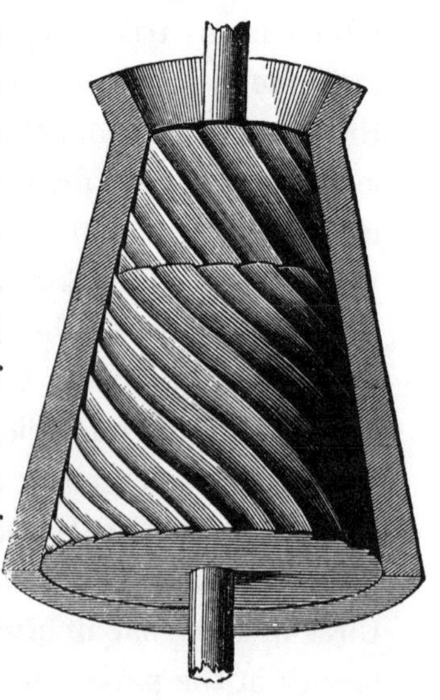

This mill was made on a principle similar to the coffee mill, but was found to exceed the simple object proposed in the specification, by grinding the grain as well as separating it from the straw. Other machines, upon the plan of rubbing out the corn, were also tried, but, on account of the damage done to the grain, were discarded. In addition to the mill invented by Winlaw, a machine was, in 1792, patented by Willoughby of Bedford, Notts, the principle of which appears to have been somewhat similar to that of Menzies. It comprised a series of loose flails, made to act upon a grated floor, and turned rapidly round by means of a horse-wheel; better, perhaps, explained by the accompanying cut, than by any other means. The straw was presented by hand to the action of the flails.

WILLOUGHBY'S THRASHING-MACHINE.

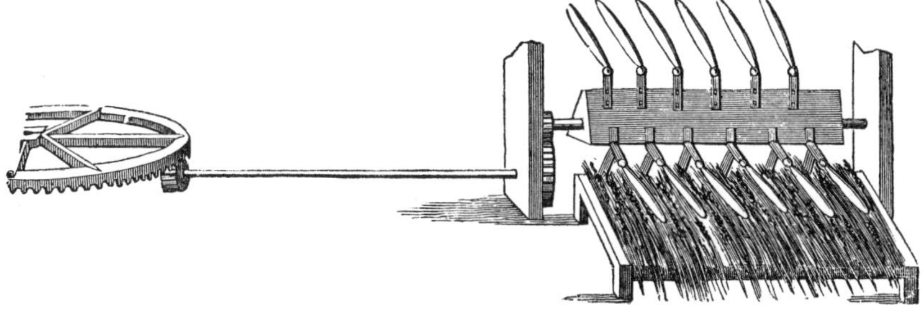

In 1795, an individual of the name of Jubb, residing at Lewes, obtained a patent for an invention of which the principal feature was the passing the straw between two rapidly-revolving beaters, under which it was held by two feeding rollers, whence the corn fell into a winnowing machine, as shown in the following figure.

JUBB'S THRASHING-MACHINE.

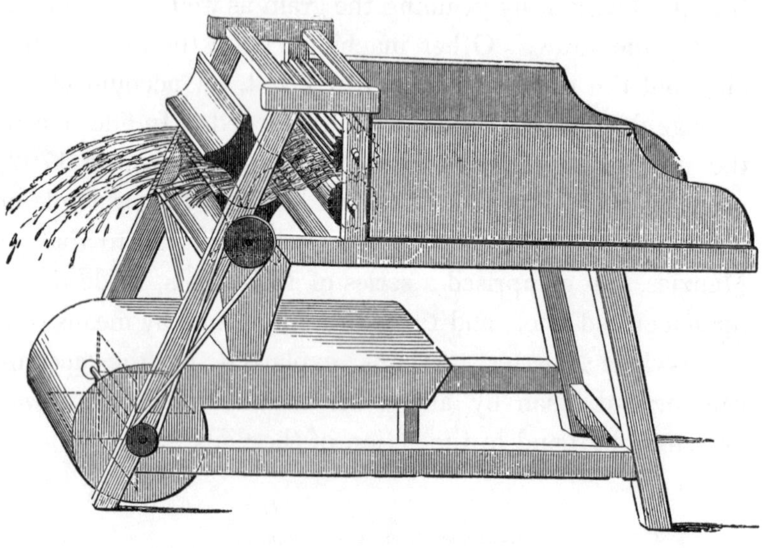

The inventive talent of our transatlantic brethren was at this time brought to bear upon this important subject. James Wardropp, of Ampthill, Virginia, invented a machine, which was introduced into England about 1796, to be worked by two men; it was made with flails or elastic rods twelve feet in length, of which twelve are attached in series, each having a spring requiring a power of twenty pounds to raise it three feet high at the point; a wallower shaft, with catches or teeth, in its revolution successively lifted each flail in alternate movements, so that three of the flails were operated upon by the whole power, viz., twenty pounds, and are seen in the accompanying cut on the point of striking, three at two-thirds raised, three one-third raised, and three at rest; consequently, the whole weight to be overcome is 120 pounds. The flails beat upon a grating, to which the corn is introduced by hand.

WARDROPP'S THRASHING-MACHINE.

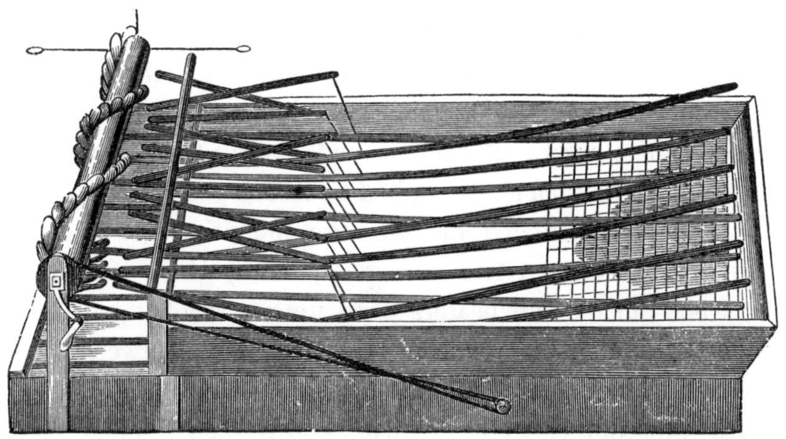

In communications from M. D. Musigny to the Society of Arts, a curious account is given of a machine worked by one horse, which, walking round a circle of forty feet, caused a cylinder, upon which thirty-two flails were placed, to revolve at the rate of twenty revolutions to one of the horse-wheel. The unthrashed straw, being laid on the circle to be described by the cylinder, must have been difficult to confine to its correct position, and awkward both to place and to remove; while the corn would be thrown about in a most slovenly manner.

Another plan, patented in 1796, by John Steedman, of Trentham, has also been exhibited, by which a number of flails fixed upon a rotatory cylinder were made continually to play upon a given spot, while a circular table, revolving horizontally, brought a fresh supply of straw under their influence.

In 1785, Andrew Meikle, of Tyningham, East Lothian,

first introduced to the public, through the medium of a gentleman of the name of Stein, of Kilbogie, the invention, the principle of which has been the basis upon which the machines in use up to the present time have been mainly constructed. It appears that, his attention having been long turned to the subject, he discovered that the plan of rubbing would never be otherwise than attended with the disadvantage before alluded to; and his son George agreed with the gentleman above named to erect a perfect machine. In 1786 he completed his intention, adopting the plan of introducing between two rollers, the corn, which was then thrashed out by four beaters fixed upon a revolving drum, each striking, as it revolved, the corn held between the rollers. The machine alluded to was erected, and found to work exceedingly well. A patent was then applied for, and after some opposition from a party not concerned in the invention, obtained. From the drawings accompanying the specification, the following wood cuts have been made, which will not be uninteresting, as proving how comparatively successful was this early design, for the full accomplishment of the purpose intended.

MEIKLE'S THRASHING-MACHINE.

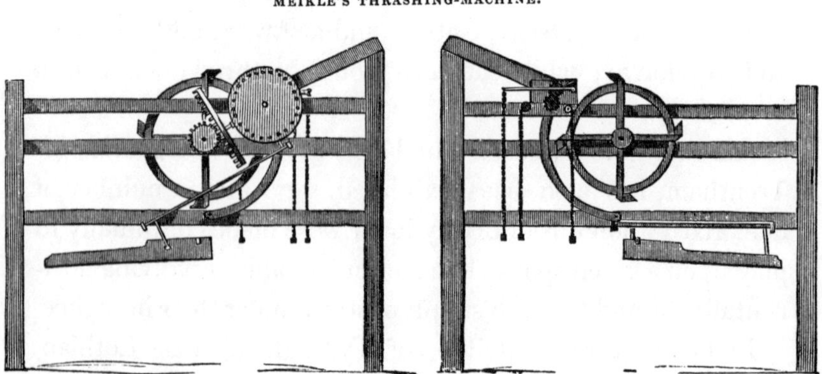

MEIKLE'S THRASHING-MACHINE.

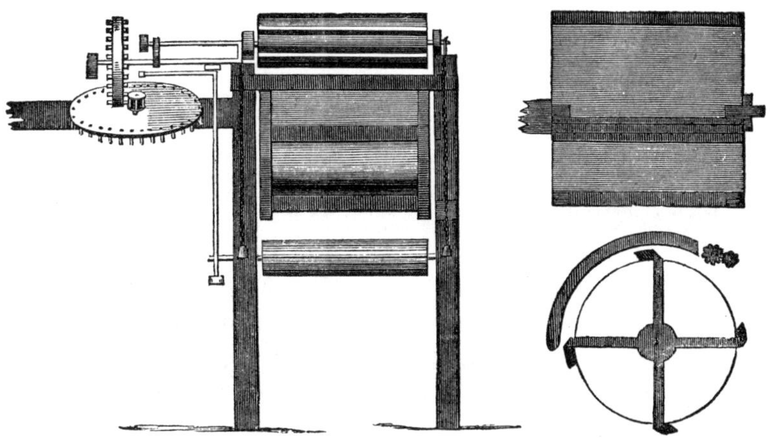

In the trials between the erection of the original machine and the obtaining the patent, a new principle appears to have suggested itself, viz., that of stripping off the corn from the ear by a comparatively sharp edge, or, as termed by him, "scutching out the grain," instead of beating it by a flat surface. The difference may be partially illustrated by supposing a handful of straw with the corn in the ear to be held in the hand, while with the flat sides of a thin strip of wood the ears should be struck or beaten; this is the operation of the common beater (see *fig.* 1). If, instead of striking the ears with the flat side, a sharp blow be given with the thin edge in the direction of the ear, it will strip the corn from such parts as the edge touches with less labour and with greater certainty. This will illustrate the scutching principle to which Meikle's beaters in his patent were applied (*fig.* 2). The difference is shown in the following cuts :—

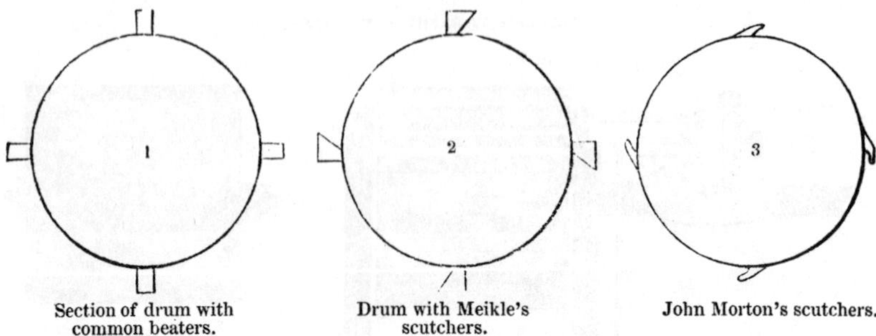

Section of drum with common beaters.

Drum with Meikle's scutchers.

John Morton's scutchers.

It is pleasing to learn, upon the authority of Sir John Sinclair, " that the inventor of this important machine was rendered comfortable in his old age, and enabled to provide for his family after his death, by the voluntary donations of his grateful countrymen." Not less honourable is the testimony of Professor Low, in his admirable treatise on the *Elements of Practical Agriculture*, that " to Andrew Meikle, beyond a question, belongs the honour of having perfected the thrashing-machine. Changes and improvements have indeed been made on certain parts of the original machine; but in all its essential parts, and in the principle of its construction, it remains as it came from the hands of its inventor."—p. 118, 3d edition.

By the drawings and specification of Meikle's machine, it appears that, at the time of taking out his patent, the plan of shaking the straw by means of circular rakes had not been suggested; and in the report drawn up for the consideration of the " Board of Agriculture" for the county of Northumberland, we find that, " in 1789, the first machine having a circular rake attached, and with fanners below, to perfect the cleaning of the grain, was erected." Although it is not there stated, we have good reason to believe that this important improvement, occasioning the addition of but one light wheel to the machine, was the invention of J. Bailey, of Chillingham, one of the gentlemen appointed to draw up the report alluded to.

An ingenious plan of yoking the horses to work in a thrashing-mill, invented by Walter Samuel, a smith at Niddry, here deserves mention. The following cut will explain the mode by which the object of making the horses pull with equal, continuous, and combined effort, is accomplished:—

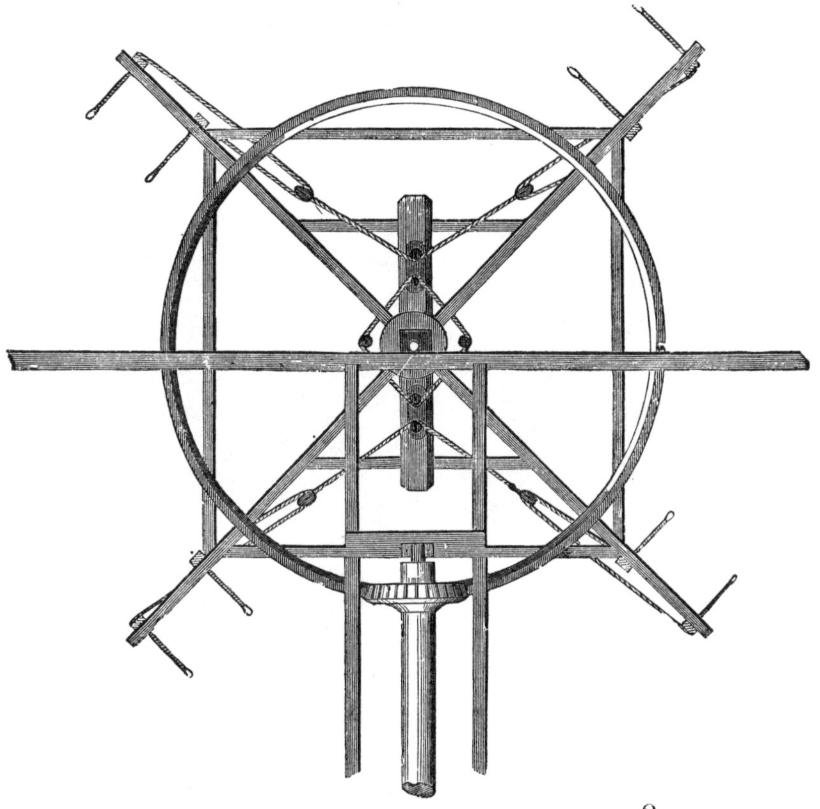

The same object has been sought to be obtained by a much more complicated contrivance, of weights acting by chains over a pulley placed in the centre of the machine, which weights were so adjusted as to be equivalent to the supposed power of the horse to whose draught they were attached, while the animal's head is tied to the shaft before him. The writer has occasionally seen these in use on the estate of the Earl of Leicester, and in various parts of the county of Norfolk; but although the horses are thus constantly kept to the collar, they have appeared, from their strained position, and from the circumstance of all the draught-irons being invariably bent, to produce a larger amount of disadvantage than was commensurate with profit.

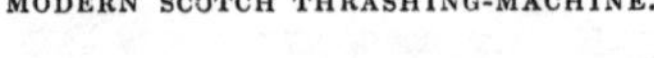

MODERN SCOTCH THRASHING-MACHINE.

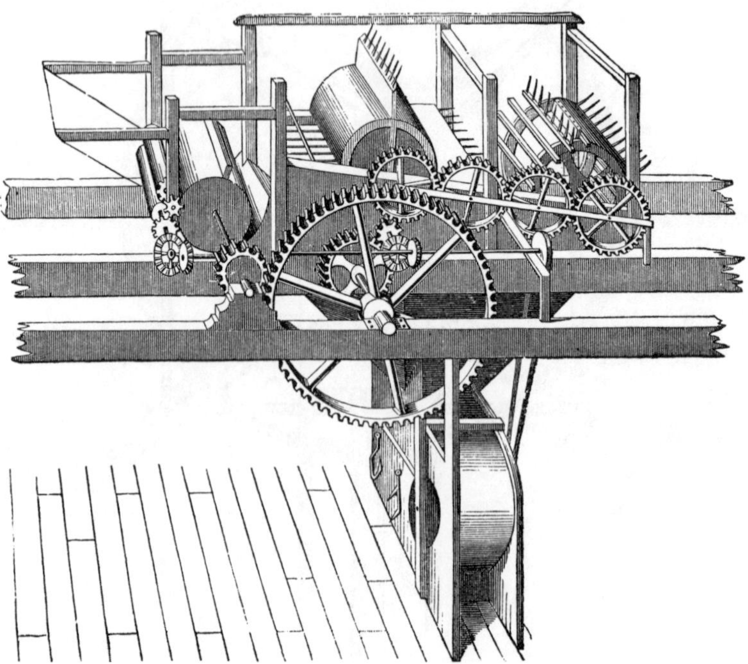

SECTION OF SCOTCH THRASHING-MACHINE.

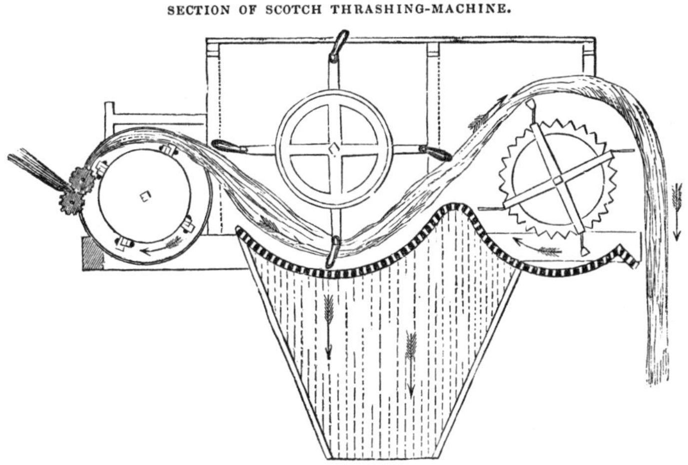

After inspecting many of the illustrations of the Scotch machine, I have selected the above drawings, as conveying the best idea of the construction and working of the instrument; for these I am indebted to Professor Low's work, from which they are copied.

In 1795, Wigfull, of Lynn, obtained a patent for an improvement in thrashing-machines, the principal feature in which was an attempt to combine the character of the impulse given by the stroke of the flail, with that of revolving beaters; his beaters, instead of being fixed (as in the case of Meikle's scutchers) on the drum, were loosely attached by means of short lengths of chain, so that the centrifugal force projected them, while in rapid motion, with increased velocity against the corn, which, passing between two rollers, was held till sufficiently thrashed: the corn was then, by means of a shaking screen, and rolling cloth or endless web, carried to the blast of a fan, where it was separated from the chaff.

This machine is gravely lauded in the *Repertory of Arts* for its extraordinary power. Alluding to one used by Ede and Nichols, of Elm, it is there stated "that it was made for four horses, but afterwards altered for six; that it cost one hundred guineas, and will, with ease, thrash from fifteen to twenty quarters of wheat, or from twenty-five to thirty-five quarters of oats in a day." A quantity which the four-horse machines of the present day would accomplish in two or three hours.

Another important step towards perfecting the thrashing-machine was next gained in one which appears to me as the first embodying the plan of feeding without the incumbrance of rollers. It was submitted by H. P. Lee, Esq., of Maidenhead Thicket, to the Society of Arts, and their gold medal was awarded him for it. A cut of this machine is here given, by which it will be seen that, with some little altera-

LEE'S IMPROVEMENTS IN THRASHING-MACHINES.

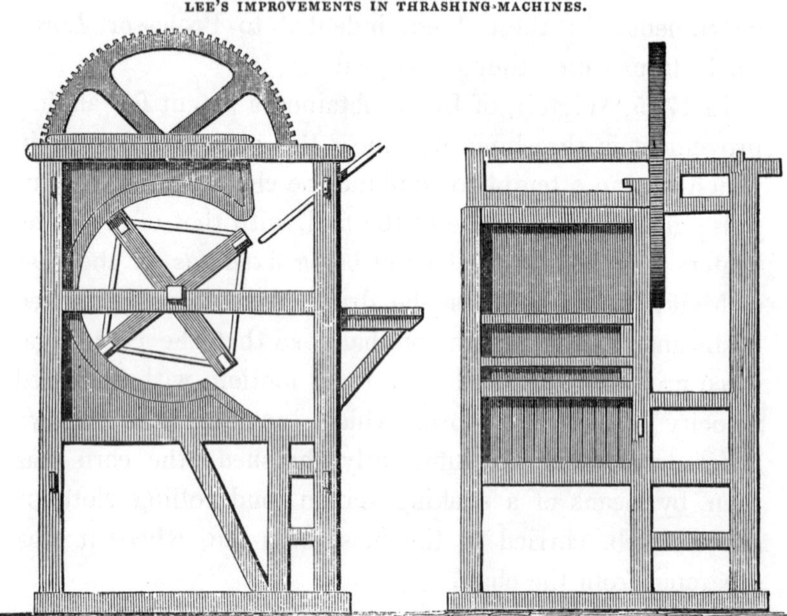

tion in the adaptation of means for gaining speed, and in the length of the concave, that this is the model upon which many of the machines in England continue to be made.

It is not possible, within the limits to which this article must necessarily be confined, to enter minutely into detail, or adequately to set forth the merits of the various inventions and improvements on this machine, for which, in the course of the last half century, no fewer than twenty-five patents have been obtained. Besides these, several medals from the Bath and West of England Society, and the Society of Arts have been awarded for other improvements, real or imaginary. But I should do injustice to the subject, did I not here mention the name of William Lester, of Paddington, as one whose mechanical talent and skill as an engineer have not a little contributed to the establishment of a higher style of excellence in this machine, as well as in other branches of agricultural mechanics than was coincident with the then spirit of the age.

The machines now in general use throughout the eastern counties of England are, with few exceptions, portable; they are frequently the property of individuals who, itinerating from farm to farm, thrash at a certain price per quarter, the farmer finding horses, and, with the exception of the proprietor, who feeds the machine, the necessary complement of men. They are simply *thrashing* instruments, having neither circular rakes nor fanners attached. The beaters, four, five, or six in number, are so placed round the drum that their beating edges radiate from the centre. These strike upon the straw, which is passed along a feeding-board placed at an inclination of about thirty degrees, tending to a point equidistant from the centre and upper part of the circumference of the drum. The *concave* or *screen* which surrounds the drum describes the third part of a circle, and is formed alternately of iron ribs and open wire-work in segments, so placed that its inner surface may be brought into near contact with the edges of the revolving beaters, and admitting of adjustment by screws to increase or diminish the distance. The usual plan is to place it within about an inch and a half space at the feeding part, and gradually to diminish the distance to an inch or three quarters of an inch at the lower end, where the straw is delivered upon a fixed *harp* or *riddle*, through which such part of the grain as is not driven through the wired part of the concave falls, while the straw is removed by forks.

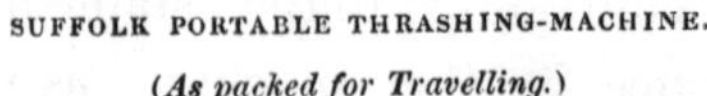
SUFFOLK PORTABLE THRASHING-MACHINE.
(*As packed for Travelling.*)

The thrashing part, commonly called the barn work, occupies a space of six feet by four and a half feet, and, together with the apparatus by which motion is communicated (which is made either for two, three, or four horses' power), may at pleasure be elevated upon a pair of wheels and axle, and thus removed by two horses.

Drawings of the machine when fixed are exhibited below.

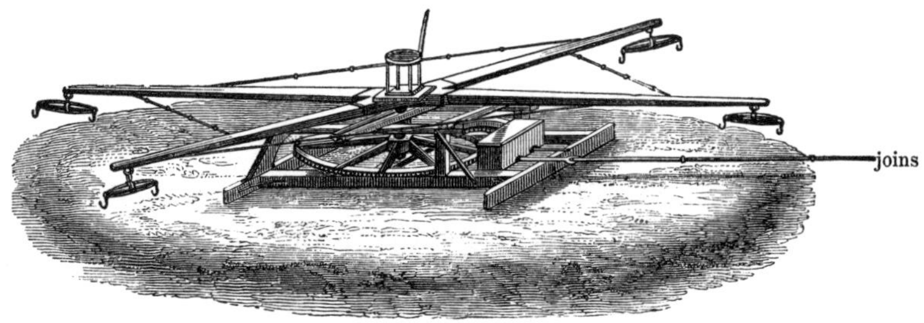

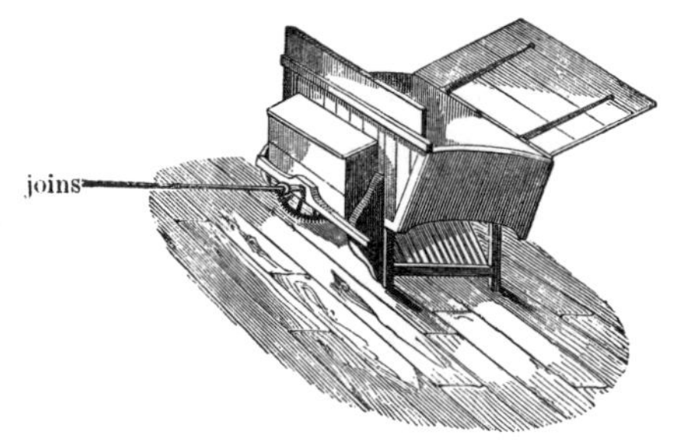

It has been urged against these machines, that they are apt to break the straw, and that they bruise and nib the barley so as to render it unfit for malting; but these faults are not so much attributable to the principles of the machines, as to the manner in which they are frequently turned out of the hands of the workman; and sometimes to the want of skill and judgment in the parties who have the management of them.

Many of these machines are made by persons who possess little claim to any mechanical knowledge, and who, purchasing the unfitted castings, by the help of village artisans, produce an imitation of those which are considered good. As the perfection of these machines must depend upon mathematical accuracy in the adjustment of all their parts, and in the truth and precision of their fittings, it is unreasonable to expect that this can be accomplished where no facilities exist beyond the forge and the work-bench; and hence arises a degree of discredit, which is unfairly thrown upon the principles upon which the machine is formed.

With these machines, properly constructed, barley may be thoroughly thrashed with as little damage as with the flail, and wheat-straw need not be so broken as materially to diminish its usefulness even for the purpose of thatching. I cannot, with Sir John Sinclair, reckon the circumstance of breaking the straw one of the *advantages* of thrashing by machinery, as it cannot be desirable that any slovenly performance of the thrashing-machine should trench upon the legitimate occupation of the chaff-engine; and I repeat the opinion that all disadvantages from the above-mentioned causes may, by a well-constructed machine and a competent manager, be entirely remedied. Where, however, as in the near neighbourhoood of London and other large towns, the straw is valuable as an article of sale, it is necessary that it should be tied in bundles or bolts, another variety of the thrashing machine, called a bolting machine, may be advantageously used. The general character of this machine is not very dissimilar to that just described, with the exception that the drum and concave are made of sufficient width to admit the sheaf lengthwise of the straw instead of presenting the ears foremost; the drum is not, as in the former case, a skeleton with beaters, but is a cylinder entirely cased with plate iron, and forms what is called a *whole drum*, upon this the beaters, eight in number, are placed longitudinally; they are formed of small strips of iron projecting not more than a quarter of an inch, the concave extends about three-fifths of the circumference of the drum, and the sheaf is introduced between two feeding rollers, as in the older machines, in order to prevent too great a quantity or too thick a wad choking and clogging the machine, which necessarily requires to be adjusted so as to leave but little space between the drum and concave.

When this machine is required to thrash beans or peas, two wooden beaters, projecting about an inch from the surface of the drum, are added, and the concave is set at a proportionate distance.

The operation of this machine, as compared with the one previously described, is slow, but its performance is very perfect, the corn is disengaged from the ear without damage, and the straw is left as straight and unbroken, and as ready to be made up into bolts, as from the flail.

I am not aware to whom the merit of the original invention of this machine belongs, but it has been much improved and brought to its present state of perfection by R. Garrett and Son, Leiston.

The latest patent which I have seen in operation is one taken out by Joseph Atkinson, of Braham Hall, Yorkshire, which appears, from the specification of a patent taken out by S. Turner, of New York, in the beginning of 1831, to be of American origin. The thrashing or beating-out process is obtained by means differing from any previously mentioned; the drum being surrounded by a series of pegs, so arranged as in its motion to pass similar rows of pegs placed at intervals in a concave, surrounding nearly one half of the circumference of the drum. This machine is not at present so fully introduced as to afford opportunity for fairly testing its comparative merits; and it would be unfair to give, upon slight evidence, an opinion which may have any tendency to increase the difficulty of the introduction of a new article. I can, therefore, say little more than that, while such trials as have fallen under my own inspection have not convinced me of its superiority, I am inclined to the belief that the principle is not so defective as to prevent its being carried out to advantage, under such modifications as may be suggested upon further trials.

I have now to draw the attention of the reader to a machine in operation upon Lord Ducie's example-farm at Whitfield, of which it is probable, in a forthcoming report of that establishment, that a full account accompanied by the necessary drawings will be given. Through the kindness of the manager, John Morton, in accordance with whose suggestions this machine was constructed, I have been favoured with an opportunity of witnessing its performance, and with the following description, which I give in his own words:—

" This machine is worked by a steam-engine of six-horse power. The corn is brought from the stack upon waggons running along a tram-road upon an inclined plane, to the doors of the building, whence, sheaf by sheaf, it is thrown by children into the buckets of an elevator, which, in its rotation, carries them to the feeding board. This feeding board is placed at a tangent from the drum parallel with its top; and, as in Lee's machine, and the portable machines in Suffolk and Norfolk, the feeding rollers are dispensed with; an endless web gradually carries the unthrashed straw to the feeding mouth, from which the revolving scutchers rapidly convey it to the concave.

" The drum and concave, being the part on which the separating of the corn or thrashing principle depends, I shall first describe:—The drum is about eighteen inches diameter, formed of sheet iron strained round a cast iron skeleton, accurately turned; upon this the beaters, or rather scutchers, formed of angle iron with its edges planed, are so placed as to describe an angle with the surface of the drum, pointing forward in the direction of its motion; these project about seven-eighths of an inch. The screen, or concave, incloses the drum to the extent of about one-third of its circumference, and consists of four or five arched pieces of grating, three inches wide, jointed together. It is made of cast iron bars, having a square section placed so that every one shall present an edge to the passage of the straw, uniting (as is not uncommon in other machines) the fluted concave of the Scotch machine with the wired grating of the English ones; the screen is supported on iron bolts, so that it approaches to within about one-eighth of an inch of the edge of the scutcher. Spiral springs surround these bolts, which permit the bars of the concave to yield when too much pressure may at any time occur between them and the revolving drum. The grain is thus separated, most of it passing through the screen of the concave; but in order that no grain shall be allowed to pass away with the straw, it is thrown upon the shaker below, which we shall presently endeavour to describe."

The shaker in this machine acts upon a principle altogether different from that generally attached to the thrashing machine: these, it will have been seen, are, in fact, revolving rakes; and of such complaint is not unreasonably made, that they not only involve loss of power equivalent to one horse at least, but a considerable portion of the grain is still carried away. To remedy this defect, by giving a motion which may more correctly be termed 'shaking,' John Morton has adopted a plan which he thus describes: — "A moveable harp, or screen, is made of spars three quarters of an inch from one another, two inches deep, three quarters of an inch wide, and six feet long; they are thirty in number, and are thus arranged over a width of three feet nine inches. These spars are fixed to two pair of frames; the odd numbers, 1, 3, 5, &c., being attached to one pair, and the even numbers, 2, 4, 6, &c., to the other pair. These frames are supported by two iron shafts, each having two cranks projecting three inches and a half on each side of them. The frames are attached to these cranks by arms with brasses, in which the cranks revolve. The shafts are connected together by a rod, so that they both move at the same time. Now, suppose them to be placed so that the cranks are one vertically over the other, and suppose the frame containing the odd-numbered spars to be attached by its arms to the upper crank, and the frame to which the even-numbered spars are fixed to be in like manner attached by its arms to the lower crank, it is very evident that by a fourth of the revolution of the shaft, the one pair of frames shall have ascended, and the other pair descended, and they will be on a level; and by the time half a revolution is completed, their relative position will be exactly reversed, the lower pair being now the upper: by the time the revolution is completed, two elevations and two depressions of the frames will have taken place."

"This up-and-down motion is exactly that calculated to shake any thing subjected to it. In the revolutions of the cranks, every thing attached to them also revolves; so that each point of the arms, frames, and spars revolves about a centre belonging to itself only. At the same time, the regularity in the length of the crank, and the uniform motion of the two shafts, has the effect of keeping the frames always parallel; their position at any one point being parallel to their position at any other."

"It is very evident, from what we have said, that any light substance, such as straw, laid upon the frame, will, by the revolution of the cranks, receive not only a shaking motion up and down, but also a progressive motion forward. The first pair of frames will raise it to the height of the crank, three and a half inches, and will at the same time carry it forward twice that distance, seven inches: its descending motion will then be abruptly stopped by the ascending pair of frames, the spars of which, rising between the spars of the first pair of frames, raise it another three and a half inches, and carry it again forward seven inches. It thus, as the revolutions proceed, not only receives a succession of blows from below, as the frames successively rise, but is at the same time carried forward from the drum, from which it is dropped on the shaker, and thus makes room for that portion of straw succeeding it."

"When two bodies with equal velocity come against one another, the collision is as violent as if one were moving with double the velocity, and the other were at rest. It thus becomes evident that the straw on the descending spars meets with as severe a blow from the spars of the ascending frame as if it were struck with twice the velocity when spread out, as we originally supposed, on a fixed horizontal platform. The effect of this succession of blows is thus more considerable than we might at first conceive, and the efficiency of the machine is proportionally increased."

"The length of the moveable harp is also of importance; for as the straw, when the cranks are three inches and a half in length, is carried forward seven inches, the number of blows it receives will depend on the length of the shaker. In the case now described, the spars being six feet long and the cranks three inches and a half, the straw will of course receive ten blows, and its progressive motion sends it off as fast as it comes from the scutcher. Each portion is removed before a second portion drops upon it."

"The motion given to the straw by this shaker is the best I have seen; the blows occasioned as each series of spars strike it from beneath, effectually remove every particle of loose grain, while the shaker rapidly carries forward the straw, and at its termination deposits it in the straw house, while the corn sifted out by its action falls before the blast of a fanner; (the construction of which is peculiar. and will be described under the head WINNOWING MACHINE;) and all the light grain and short straws thrown out by the first winnowing into the light corn spout is then taken up by another elevator, deposited again upon the feeding board, and passed a second time through the drum, in order effectually to separate any that may remain. After passing through another winnower, the thoroughly cleaned corn is taken up by a third elevator and dropped into a hopper, through which it passes into a sack, which is placed on a weighing machine, and it is there weighed and left thoroughly fit for market.

I have introduced at length the description of this machine, or rather series of machines, as being the most complete of any that has fallen within the range of my observation; the most comprehensive in its design, performing every operation, from receiving the sheaf at the barn door to depositing its grain in a clean state weighed up in the sacks, and excellent in the greater part of its detail, which is carried out, both as regards ingenuity and workmanship, in a style very superior to the general character of agricultural machines. I would, however, remark upon one part of its operations, which I am inclined to consider superfluous; or rather, if there be a necessity for its adoption, it argues a degree of imperfection in the working of the drum with scutching beaters, which, in a drum constructed with the common beaters, I conceive, need not exist:—I allude to the second elevator, whose sole business appears to be to carry from the light corn spout of the first winnower the light corn and short straw to be re-passed through the drum, for the purpose of separating any grain that may remain on the heads after they have once passed the drum. I question whether, if a perfect separation of the corn from the straw could be effected in its passage between the drum and concave (which I conceive may and ought to be accomplished), there would be any advantage in a second fan, which then, together with the second elevator, might be dispensed with.

This brings us to the consideration of the comparative advantages of the scutching and the beating principles. I am willing to believe, as in the illustration of the thin strip of wood used edgewise, a much lighter blow is effective to clear out the corn from every ear which the edge of the strip or scutcher shall pass; but it will be obvious it will only act on these, and consequently it will be necessary to bring the edges of the scutchers so near to the concave as not to allow more than the thickness of one ear between them; hence the necessity of the drum, as described above, revolving within one-eighth of an inch of the concave. The tendency of the scutching principle, in its attempt to carry them rapidly through the narrow interval, will be occasionally to break off some of the ears; and these ears, when no longer supported by the straw, may fall behind the scutchers, and thus be carried through unthrashed. The centrifugal force would no doubt have a tendency to throw them again to the action of the next bar of the concave; but it must be borne in mind, this part will be pre-occupied with the straw which is in the act of passing, and not until the straw is released will it have opportunity to fall; it must then fall with the straw, and, but for the perfect operation of the shakers, be carried away with it. The shaker, however, remedies this disadvantage; it therefore falls into the winnower, and, being lighter in proportion to its bulk than the corn, is blown with the chaff into the light corn spout; and this involves, we suppose, the necessity of the second elevator, whose business it is continually to remind the scutchers of any defect in their first performance. I have some fear that the same cause which allows the ear to pass unthrashed the first time may admit of its doing so the second time; and that, not having any longer the straw to hold it to the operation of the scutcher, a single ear may thus traverse the drum, the fanner, and the elevator, a second or even a third time, without the corn being perfectly separated from it.

Having described the operation of the scutching or stripping principle, I have now to consider that of the beaters; and here I should take objection to the statement, that by the blow with the flat side of the strip of wood this is adequately illustrated, inasmuch as a blow with such an instrument cannot be given with sufficient force to produce the effect intended by the machine, and which, to be effective, must be severe. The principle by which the beating process operates, depends upon the comparative velocity with which bodies of different specific gravity meet atmospheric resistance, when impulse has been communicated to them. This may be illustrated by wrapping a leaden bullet in thin paper; to this if a sharp blow be applied, the bullet will fly to a considerable distance, while the lighter covering will be left. If one hundred bullets so wrapped up, and in a cluster, were struck with sufficient velocity, the effect would be produced on all. I admit the power by which this impulse is communicated must be increased in the ratio of their number; but this is not difficult to accomplish, and is here a question of power, not of principle. But if one hundred of these bullets in cluster be subjected to the scutching or stripping operation, while it may be freely granted that those immediately coming in contact with the sharp edge will be stripped at less expense of power, the operation will need to be repeated again and again, until each has come into contact with the edge, or with an unyielding surface in each other; the husks of corn, being yielding surfaces, will retard the perfect operation of the stripping process, unless ultimately forced through an interval not more than equal to the bulk of each. The beaters, on the contrary, do not require to force the corn through or between a narrow interval, as they need only to be so far surrounded as to confine the straw, which otherwise, by centrifugal force, would be drawn out of the range of the beaters. The question is, therefore, narrowed to the consideration of whether the power required to communicate the impulse by the beaters is greater or less than that which is necessary to strip the ear; it being borne in mind that the stripping process has to be communicated to each ear, to force it through a narrow interval, and that the friction in this operation is not limited to the ear, but to all the straw which has subsequently to pass through the same narrow space.

From a series of experiments, which I have, through the kindness of practical agriculturists, been attempting, I am of opinion, that although the operation of scutching may be performed at less expense of power, if partially executed, and that a greater quantity of straw may pass the machine, and consequently more corn may be obtained in a given time, yet, unless the scutchers are placed very near the concave, the work will be imperfectly performed, and much of the corn will remain in the straw, and escape. But, if placed so near as is requisite for clean thrashing, the liability to break off the ears will be increased, which will involve the need of re-thrashing; and the friction, from the narrowness of the interval, will then require, to the full, as great power as the beaters.

Every one acquainted with this subject is well aware of the difficulty with which trials of this implement are necessarily attended; the labour and loss of time attaching to putting up and taking down heavy instruments to fix others in their place (which, if the experiment is to be fairly tried with grain of the same quality and in the same condition, cannot be dispensed with), the difficulty of getting men equally competent to manage them, and the trouble occasioned to the manager or owner, are so many drawbacks upon the facility with which sufficient evidence can be obtained to guide to sound conclusions on the respective merits of the varieties of these machines. And to this must be added that, under certain circumstances, machines, which will do their work admirably when every thing may be favourable for them,—such as the corn in good condition and the straw not of too great length,—are sometimes found, under adverse circumstances, to do their work very incompetently; while the slower and heavier-working machines, which in the first instance could not keep pace with them, are found efficient in the latter. I instance this to show, that with all the care and pains I may have taken, and it has not been little, it is possible that comparative trials, under other circumstances and in different hands, may not lead to the same results. The impression, however, produced on my mind, by a careful weighing and balancing of evidence, is—that for heavy corn, loose in the husk, wheat, peas, and beans, the beating principle is decidedly the best, the most perfect, and obtained at least cost. For barley and oats the scutching principle is more adapted. But I have still considerable doubt whether, even for these, the operation is so perfect as with the beaters. My experiments, it should in fairness be stated, were made with scutchers on a drum only seventeen inches in diameter, being about the same size as the drum with beaters; and it is not impossible that the scutching principle may be used to better effect on a drum of larger diameter. I have thus expressed my own impressions, and trust that I have said sufficient to induce the comparison to be made by the experience and observation of others, the result of which I shall be glad to learn. Should further trials prove that either the scutching or beating principle may be used, under certain circumstances, to greater advantage than the other, it would not be difficult so to construct a drum as that the change should be readily effected.

Having thus briefly described the machine as regards its thrashing operation, it will be expected we should address our attention to the moving power, and consider which is most advantageous in connexion with what is most readily available. The uncertainty of wind renders this power, if for an object to be performed at certain times, comparatively of little value; and, as one prominent advantage of this machine lies in its ability to do a great quantity of work on an emergency, we may conclude it is less available as a general principle than either water, steam, horses, or manual labour.

Where the locality admits of the use of a water-wheel, this power is most economical and easily managed; but the advantage is limited to peculiar situations.

Where the quantity of work to be performed is sufficient to repay the interest of outlay, expense of wear and tear, &c., a steam engine would be most advantageously employed on the farm. Of its economy, as compared with either horse or manual labour, there need be no question. But as few farms in this kingdom at present have these appendages, the question for consideration is narrowed to the comparison between horses and manual labour. On the authority of Dr. Gregory, the dynamic power of a horse at a dead pull may be calculated in the main as equivalent to that of six men, or to 420 lbs., if exerted in a direct line; but the result of experiments made by Tredgold tends to prove that sustained effort at the rate of three miles per hour must not be calculated at more than equivalent to 120 lbs. drawn over a pulley. This, taking six hours of labour per diem, as the utmost he would recommend, would be the maximum of useful effect. Under the circumstance of any deviation from a straight line, this must be materially reduced; and in describing a circle of eighteen feet diameter, the cramped position of the horse will probably prevent his power from being advantageously exerted to the extent of much more than half. It will, therefore, be seen that a very large proportion of dynamic effort is wasted; and this not only arises from the constrained position of the horse's movement, but from the friction of the mill by means of which motion is communicated to a machine.

It is affirmed by Emerson that a man of ordinary strength, turning a roller by the handle, can act for a whole day against a resistance equal to 30 lbs. weight; and if he works ten hours a day, he will raise this weight through three feet and a half in a second, or about two miles and a half per hour.

Animal power is, however, so varied by the character of the exertion, that it is difficult to arrive at a correct calculation. The late Robertson Buchanan ascertained that in the action of working a pump, of turning a winch, of ringing a bell, or rowing a boat, the dynamic results were respectively as the numbers 100, 167, 227, and 248.

Having caused a machine with beaters to be constructed,

worked by four men whose force should be exerted as in the manner of rowing a boat, the results, as compared with a machine requiring the force of four horses in a circle of eighteen feet diameter, I found might be taken on an average as five to twelve. It was thought the continuous effort might be for an equal length of time exerted by six men relieving each other at intervals, as by the same number of horses relieved in the same way.

Subsequent experiments have, however, led me to believe that the manual power was somewhat over-estimated as to its capability of endurance; but I instance this merely to show that, although advantage may be on the side of horse labour for large quantities, manual force is not so inapplicable to this object as most writers have represented

RANSOME'S HAND-THRASHING-MACHINE.

it to be; and I am of opinion that, on small farms, hand machines may with great advantage be used. The accompanying cut exhibits a simple and effective hand thrashing-machine, which was exhibited at the Royal Agricultural Society's meeting at Liverpool, and obtained the commendation of the judges (*vide* their report). It is worked by four men, and the moving power being obtained by means of a lever on the one side, and by a crank handle on the other, the men working it may relieve each other by change of motion. It requires one man to feed the machine, and the number of hands necessary to bring the sheaves and remove the straw will depend upon the distance it has to be conveyed. When the straw is short, and the wheat of average yield and in good condition, it will thrash at the rate of ten to twelve bushels per hour; but with mown wheat, or wheat which is long in the straw, the quantity will be proportionally less Machines on this principle are sometimes attached to a set of horse works, for one or two horses, and are found to answer well.

The cost of the thrashing-machines of former days varied considerably, and their performances were very unequal. It may not be uninteresting to give a list of some of these, in order to afford the opportunity for comparison with those of the present day. The following are extracted from the Agricultural Reports:—

	Thrashing per Day.
In the reports of Roxburgh and Selkirk, in 1796, R. Douglas states that mills by water, or with 4 horses, would do great execution.	25 or 30 bolls, or from 150 to 180 bushels.

In the report of Norfolk, in 1804, Arthur Young gives an account of machines which belonged to the following parties:—

	Thrashing per Day.
Droziers, Reedham, built by Wigfull, cost 120*l.*, worked by 7 persons and 6 horses.	40 co. wheat, or 50 co. barley, or 60 co. oats or peas.
Farrow, Shipdam, built by Wigfull, worked by 7 persons, and by 4, 5, or 6 horses.	20 co. wheat, or 30 co. barley, or 40 co. oats or peas.
Beck, Castle Rising, built by Wigfull, cost 200 guineas, worked by 6 persons, and 4, 5, or 6 horses.	32 co. wheat, or 64 co. barley, or 80 co. peas.
Whiting, Tring, built by Fordyce from Scotland, cost 200*l.*, worked by 6 persons and 6 horses.	24 co. wheat, or 55 co. barley, or 63 to 84 co. oats.
Bevan, Riddlesworth, built by an engineer from Leith, cost 100*l.*, worked by 10 men, and 8 horses.	40 co. wheat, or 40 co. barley, or 50 co. oats,
Coke, Holkham, cost 600*l.*, worked by 12 men and 8 horses.	64 co. wheat.
Reeves, Heverland, built by Assby, Blyboro', cost 100 guineas, worked with 2 or 3 horses.	30 co. wheat, or 32 co. barley, or 40 co. peas.
Styleman, Smithsham, cost 300*l.*, worked by 10 persons and 8 horses.	80 co. wheat, or 120 co. barley, peas, or oats.
In the report of Kent, R. Boys, in 1805, remarks on the only thrashing-mill then in Kent, which, by a number of improvements, and after many alterations, he finds to answer extremely well; and he states that it requires 4 horses and 12 men to work it.	24 qrs. wheat, or 32 qrs. barley, or 40 qrs. oats.
In Sir John Sinclair's *System of Husbandry*, published in 1812, we find an account of R. Kerr's machine, which, with 6 horses, 4 men, and 4 women, would thrash	50 bolls, or about 300 bushels of wheat,

Considerable improvements have since been effected. In the statements of the trials of implements at the Royal English Agricultural Society's meeting at Cambridge, in 1841, the quantity of wheat thrashed by two four-horse portable machines manufactured by J. R. and A. Ransome, of Ipswich, and R. Garrett and Son, Leiston, was respectively sixty-one bushels and three quarters of a peck, and sixty-one bushels and a quarter of a peck; and the corn was clean-thrashed and uninjured.

This must not be taken as a criterion on which to found an average, as it was doubtless the result of stimulated exertion; but it is not unusual with machines of this construction, with reaped wheat in fair condition, to thrash 50 quarters or 400 bushels in a day of ten hours, and the same quantity of mown barley.

It should, however, be observed, that these, having neither rakes nor fans, the work of which is done by hand, would require eight men and five boys, and a change of horses in the day.

SECTION IX.

THE WINNOWING MACHINE.

Having traced the excellent modern implements commonly used for raising the Farmer's crops, and separating the seed from the straw, the next valuable instrument which demands attention is the Winnowing Machine, for separating the corn from the chaff.

The earlier mode of producing this effect was by a very simple blower, or fan, composed of four pieces of wood, placed longitudinally in the direction of an axis, upon each of which was fastened a piece of cloth or canvas; as this was made to revolve by turning a handle or winch, it produced an artificial current of air, before which the corn was thrown by shovels, so that in passing the blast, the lighter chaff and dust was carried away; and amongst the labourers this rude instrument is still in use for dressing their gleaning corn.

The machine now in general use, is a partially enclosed box containing at one end a chamber in which a fan is caused by means of toothed wheels to revolve very rapidly, forcing a strong current of air in the required direction. The corn being put into a hopper at the top, descends upon sieves arranged in front of the fan, and during its descent is subjected to the action of the blast, by which the chaff and other light matter is separated and blown away; the sieve is kept in constant and rapid motion by means of a connecting rod attached to a crank on the axis of the fan," and the corn, being violently shaken whilst still exposed to the stream of air, is driven to the edge of the sieve and falls upon a long inclined screen, whilst the small corn, seeds, and remaining dust pass through the sieve, and the corn, when it has reached the bottom of the screen falls out perfectly cleaned and fit for market.

Various are the accounts given of the introduction of this machine, and many the claimants for the credit of having been the first maker of this piece of mechanism in England or Scotland. All, however, agree that the idea, design, or model was originally furnished from Holland; earlier, however, than the date of any of these claims by at least a period of twenty years, we learn from the papers of Robert Somerville of Haddington, that in 1710, pursuant to articles of agreement between himself and Fletcher, Laird of Saltoun, James Meikle (father to Meikle of thrashing-machine memory) visited Holland for the purpose of learning "the perfect art of sheeling barley," in order to the introduction of the barley mill. The same authority, writing in 1805, states, "that on Meikle's return he made the first fanners which were seen in Britain;" and that these were in use only a few years before that date at the Saltoun barley mills. That the machine was not made public till many years after its first introduction, may be attributed to a clause in the above-mentioned agreement, by which Meikle was bound, on leaving Saltoun's service, "not to profit any more by this mill, nor communicate the arts he had learned to any other." In 1737, through the medium of Rogers of Cavers and others, it was brought into more general use; and in 1768, A. and R. Meikle obtained a patent for a machine of this kind. Although a very considerable advantage over the plan of dressing by hand, these machines still appear to have been but very imperfect. the corn having to be passed through them twice or thrice in order to be perfectly separated. And in 1798, R. Douglas, in his *Agricultural Survey of Roxburghshire*, remarking upon these defects, mentions an improvement invented by one Moodie of Lilliesheaf, " in which he had happily combined some properties of other fans, so that the grain at one operation could be both separated from the chaff and lighter seeds, and completely riddled of all sorts of refuse."

COOCH'S WINNOWING-MACHINE.

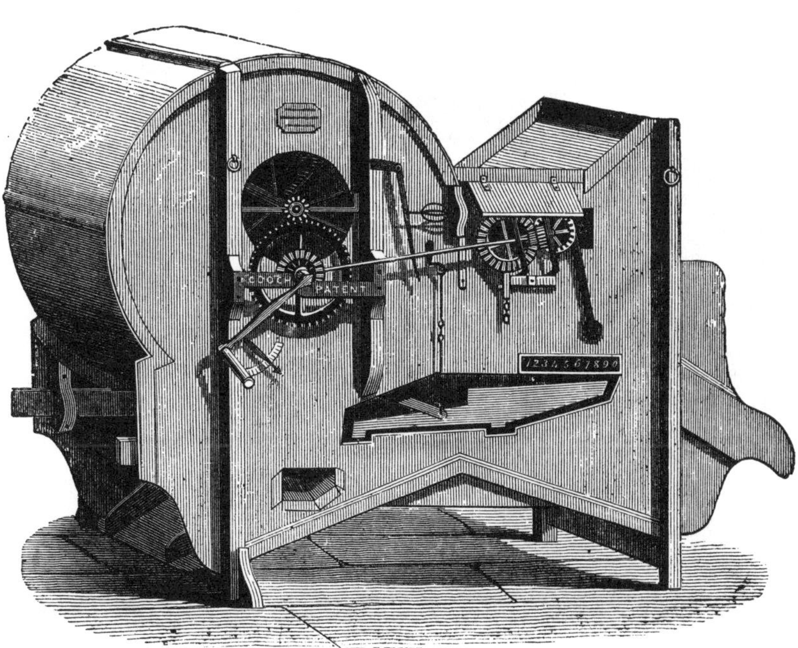

Other patents had been taken out which do not appear to have involved much real improvement, till, in 1800, J. Cooch, of Northampton, patented the machine which has since been known by his name, and has obtained deserved commendation, being in use and approved beyond most at the present day. This machine dresses all kinds of seeds, and its work is performed in a perfect manner; its principle involved more mechanical combinations than were at that time generally understood by the class for whose use it was intended; and this, together with its then cost, long retarded its more general adoption. I do not know of any better machine in general use; in proof of the estimation it deservedly continues to hold, the R. A. S. E. prize was awarded to it at the Liverpool meeting, 1841.

ELMY'S WINNOWING-MACHINE.

In 1812, John Elmey obtained a patent for improvements in winnowing-machines, and produced a very efficient implement; the arrangement of its various parts was simple, and greater effect was obtained from the blast. Comparing this with the drawings and description of one I find in the *Edinburgh Journal of Agriculture*, and with that described by Professor Low, I have little doubt of their general identity with this, and as it is the model upon which the machines in general use are now made, I subjoin a sketch of it. It is very simple, easy to work, and not likely to be out of order, and is in very general use.

SALTER'S WINNOWING-MACHINE; SIDE ELEVATION WITH FRONT REMOVED.

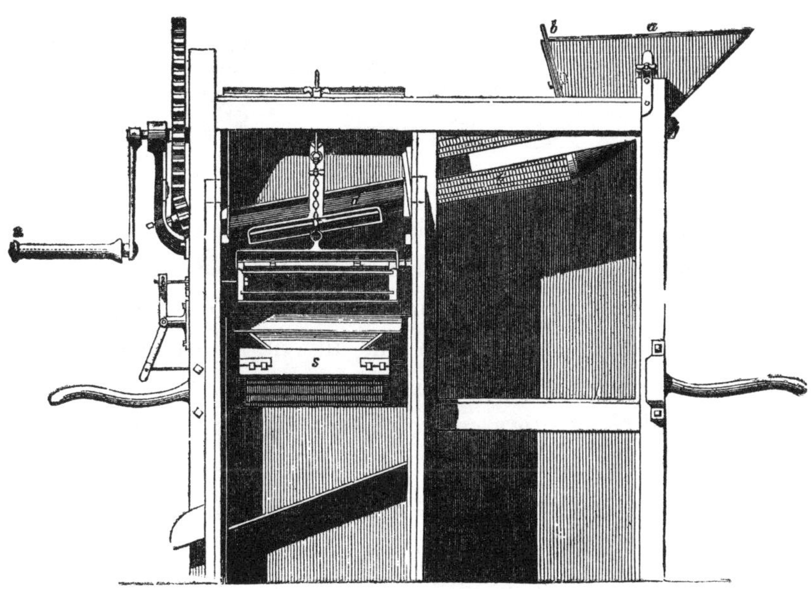

SALTER'S WINNOWING-MACHINE; END ELEVATION.

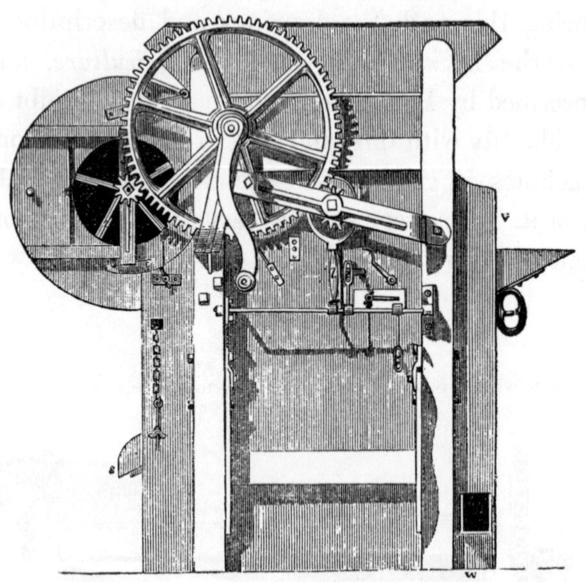

In 1839, T. F. Salter obtained a patent for a machine for winnowing and dressing corn and seeds, which at the R.A.S.E. meeting at Cambridge was exhibited, and obtained the silver medal. In this invention are combined the principles of the hummelling machine, described at the close of this section, with the operations of the common winnower.

The undressed grain from the hopper passes through a cylindrical sieve, having within it a rotatory spindle, upon which short blunt arms are arranged in a spiral direction; these agitate the grain as it passes along, and thus separate the small dirt and dust as well as the awns of barley, which fall through in a closed box or cupboard. The cylinder is placed in a slanting direction, and is provided at each end with slides, which regulate the quantity and speed with which the grain shall pass. Through the slide aperture at the lower end, the grain is introduced upon other sieves, which, having a backward and forward motion, distribute it equally over their surface while it is subjected to the blast of the fan, driving obliquely through the sieves; this carries the chaff out of the machine; the grain falls on a screen, which, having a similar motion to the sieves, separates from it all small seeds, and the dross corn is carried away in a division formed for the purpose. The grain, dross corn, and chaff are thus all thoroughly separated from each other, and the dust, dirt, and small seeds, having fallen in an enclosed box from the cylinder, may be entirely removed.

I have heard this machine highly approved by many, and when pains are taken to separate the corn from the short straw, &c., previously to submitting it to the machine, I believe it to be very effective; but as there is some degree of complication in its details, it is chiefly suited to those to whom a high degree of excellence in the manner of "making up their corn" is a matter of more importance than the time or labour it may require.

I now come to the description of the winnower used in combination with the thrashing apparatus at Whitfield, in which the principal feature is the improvement of the fan or blower. Having noticed that the ordinary form and position of the fans, which are flat boards, radiating from the centre as seen in the drawing, tended to keep the air constantly whirling within the casing, rather than to force it forwards; and that if, instead of being flat, they were curved forward in the direction of their motion, they would draw the air in from the tube, and force it out at the sides, R. Clyburn, of Uley, the engineer by whom the machinery at Whitfield was executed, constructed a blower, in which by a backward curvature of the fans, and a different arrangement of the chamber in which they revolve, the tendency to form a vacuum is considerably increased, and greater force is consequently obtained from the blast. The arrangement of the chamber in which the fans are made rapidly to revolve will be better understood by the following sketches, the first showing imaginary spaces required for the passage of the air drawn off by each fan, and the second the position of the fan in the eccentric box which surrounds it.

DIAGRAMS.

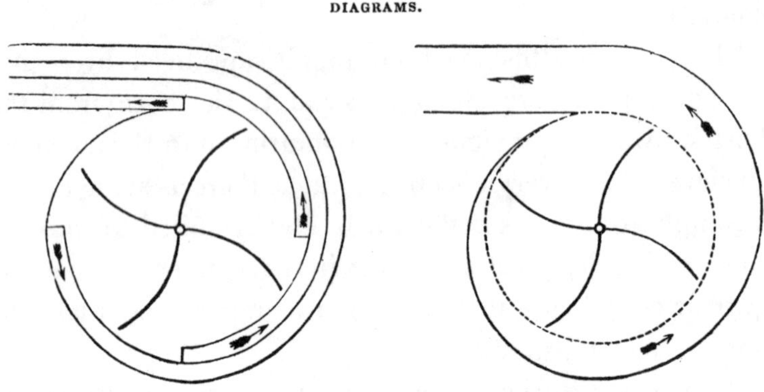

We are not disposed to leave this part of our subject without some allusion to an invention for still further carrying out the process of cleaning corn, known as Tuxford's Reeing-machine. This consists of a series of sieves, to which a rotatory motion is given : the grain is by this means separated from any small dust and dirt which passes through the wires of the sieve, while all the lighter rubbish is by the motion brought to the top, whence it is removed by hand. This implement is more, perhaps, adapted for millers; and its cost presents, in its present form, a bar to its general introduction. If it could be reduced to the power of being worked by hand, it would be a very valuable assistant to the proper preparation of the grain for the market.

BARLEY HUMMELLING MACHINE.

This instrument, though not a winnowing-machine, deserves, as a machine for preparing barley for the market, notice here.

It consists of an inclined cylinder of wove wire, similar

in form to that of a flour-dressing machine, but of much coarser texture, within and concentric to the cylinder is placed an axis on which is fixed a series of knives set in a spiral form, and nearly reaching to the wire. The cylinder is fixed in a box, on the top of which is set the hopper, into which the barley is to be thrown, and at the bottom is a spout whence it is discharged when cleaned. The shaft with the knives is driven at a rapid rate by means of toothed wheels, and the barley contained within the cylinder is violently agitated, by which the awns of the barley are broken off and driven through the interstices of wire work, and are deposited in the bottom of the box.

The invention of this machine is claimed by several, but in *British Husbandry*, vol. ii. p. 204, its present simple form is ascribed to an individual of the name of Grant, of Granton, in Aberdeenshire. I do not know whether this was prior to the invention of Mitchell's, described in No. 7 of the Appendix to Sir John Sinclair's Systems of Husbandry, but it is a much more simple and effective instrument.

The one from which the above description and cut was taken, was furnished me by R. Garrett, of Leiston, who has, I believe, been instrumental in its further improvement.

SECTION X.

CHAFF ENGINES.

I CAN find no early traces of the construction of chaff engines, although it is probable that, in some rude form or other, these existed from a very remote period. Some of the early agricultural writers direct the farmer, to give their cattle cut hay and straw; thus M. P. Cato, the earliest of agricultural writers, lib. 54, mentions chaff as the food for oxen, with the ordinary provender of the modern farm yard, and directs these to be given mixed with salt. The leaves of the elm, the poplar, the oak, and even of the ivy, appear to have been extensively used as food for cattle by the Italian farmer of those days, and it is difficult to imagine how this could be done to any extent, except by means more powerful than the unassisted knife or the chopper. The advantage of confining in a trough the fodder intended to be cut, would soon be apparent to the workman, and hence would readily originate the old-fashioned, and extensively diffused hand chaff-box.

I am not aware of any attempt to improve upon the plan of pressing the hay in a trough, and bringing it in small portions to the front edge, where it was severed by a long knife attached to the end of a lever, till in 1794-5

J. Cooke, clerk, of London, and W. Naylor, of Langstock, respectively obtained patents for machines for expediting the process.

SALMON'S STRAW-CUTTER.

In the year 1797, Robert Salmon, of Woburn, whose inventive talent and practical experience added many and various original ideas and improvements to the then limited knowledge of agricultural mechanics, constructed a chaff engine, which, although cumbrous in its appearance, was effective in its operation, and furnished the original idea, which was subsequently improved upon, first, by Rowntree, and afterwards by Thomas Passmore of Doncaster, the latter of whom, in 1804, patented the machine known as the Doncaster engine, upon the plan of which, for many years, most of the engines in the midland and eastern counties were made, and even at the present time, few of the machines in general use are found more effective. A reward of thirty guineas was conferred on Salmon by the Society for the encouragement of Arts, &c., for this improved machine.

The plan of Salmon's engine, as exhibited in the accompanying illustration, may be described as the fellies of two wheels connected together, and knives fixed upon them, the edges of which are fixed at an angle of 45° from the plane of the wheel's motion. Springs are fixed on the wheels thus connected, by means of which the knives are pressed forward against the box. On the other side of the knives, wedges are fixed to counteract the pressure of the springs, should it be too great. To a circular block of wood, having four holes, and fixed on the wheel, one end of the feeding arm is screwed, and is fixed to the cross bar by a pin, moveable at pleasure to five different holes, by which arrangement twenty different changes of length of chaff may be obtained. Two spiked rollers in the box are turned from the outside by ratchet wheels, so that the straw is at rest during the time the knife is cutting upon it. A weight is suspended by a lever under the box, which will assist in forcing the straw forward, and counterbalance the rachet wheel of the upper roller. Equal pressure is given to the straw by a chain passing from near the fulcrum of the lever to a roller with two small bars of iron, which are attached also to the projecting axle of the upper feeding roller.

PASSMORE'S CHAFF-ENGINE.

Passmore's machine, it will be perceived, was very similar, but its mechanical combinations are advantageously simplified, and the circle upon which the knives are fixed is much reduced.

In 1800 and 1801, W. Lester, of Paddington, patented a straw-cutter, which with some alterations, is much used at the present day, and is known as the "Lester engine." It is a very simple machine, having but one knife placed on a fly-wheel.

IMPROVED CHAFF-ENGINE, ON LESTER'S PRINCIPLE.

The fly-wheel turns on a cranked spindle, which communicates motion to a rachet wheel, fixed at the end of one of the feeding rollers by means of a small hook or catch, which is capable of being so adjusted as to lift one two, three, or four teeth at each revolution, and by this is regulated the length of the straw projected in front of the face plate, and which is severed by the knife. On the roller was fixed a revolving cloth or endless web, which passed over another roller at the hinder end of the box. A heavy block was used to compress the straw. In the more modern engines the rolling cloth is entirely dispensed with, as the purpose for which it was intended is effected by the introduction of an upper feeding roller, to which motion is communicated by a pair of cog-wheels, one of which is attached to the lower feeding roller before described; the heavy block is substituted by a pressing piece, which, receiving its motion from the cranked spindle, alternately presses down the straw previous to the cut, and rises afterwards to allow the straw free passage. A cut of the improved machine is given above; it is made of different sizes, and the larger are frequently used with horse-power.

This is one of the best modern chaff-engines, it will adjust and vary the work to the following lengths of cut:— ¼ inch, ½ inch, and ¾ inch.

	Bushels of fodder per hour.
At ¼ inch it will cut from .	18 to 20
½ ,, .	40 to 50
¾ ,, .	50 to 60

Another chaff-cutter is made on the same principle, but a size smaller, which

At ¼ inch will cut from .	10 to 12
½ ,, .	30 to 40
¾ ,, .	40 to 50

Passing by several which, in the course of the next fifteen years, were introduced, but which, however ingenious, were too complicated and cumbrous for general use, in 1818, we find a simple invention was patented by Thomas Heppenstall, of Doncaster. It consisted in the application of a worm to turn two wheels, which in their revolution meet each other. These wheels are attached to two feeding rollers, which convey the straw forwards to the knives.

Two of these knives are placed on a fly-wheel, which is fixed upon the same spindle as the worm. This is the simplest form of chaff-engine, and with a slight alteration, substituting wheels with the cogs on the face instead of on the outer edge, is the common form for the small engines now in use.

This engine is suited to gentlemen's stables and small establishments, and being entirely of metal, is adapted for hot climates. It will cut from 15 to 20 bushels of fodder per hour.

HEPPENSTALL'S CHAFF-ENGINE.

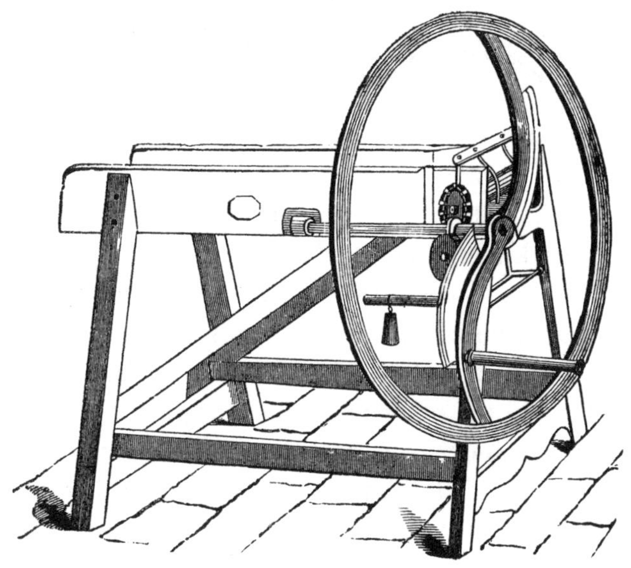

Two patents have also, within the last year or two, been taken out for considerable improvements on the chaff engine, one by Lord Ducie, in connexion with Clyburn and Budding, two engineers residing at Uley.

LORD DUCIE'S CHAFF-CUTTER.

Lord Ducie's machine is thus described in the Judges' Report of the Implements shown at Liverpool, R. A. S. E. Journal, vol. ii. p. 111. It is stated that, upon trial, it performed its work admirably. "The cutters consist of two series of thin blades or knives, with serrated edges, coiled spirally round a horizontal rotating cylinder, and presenting their edges at an angle to it. The one series is coiled from left to right, and the other from right to left, meeting in the middle of the cylinder; an unbroken continuity of cutting action is thus attained. A pair of feed-rollers is driven from the spindle of the cutting cylinder, which again gives motion to an endless cloth, upon which the material to be cut is placed, and by which the supply is maintained. The speed of the feeding-rollers is regulated by a highly ingenious and simple application of the worm and wheel. The wheel fixed on the roller is so constructed, as to admit of being driven by worms, with threads varying from one to four;

thus, by changing the worm on the axis of the cylinder, (which is also accomplished in a dexterously mechanical manner) the hay or straw is cut into lengths of from a quarter of an inch to one inch. This machine may be worked by manual, animal, or steam power with equal convenience."

RANSOME AND MAY'S CHAFF-ENGINE.

The chaff-engine, patented by C. May, is a successful attempt at combining the advantages of some of the older plans, with the power of altering the length of the cut, and also avoiding the difficulty of supplying the material to be cut, so evenly, that it may be delivered at the mouth pressed so close, as to stand against the knife. The alteration of the length is accomplished by adding a second shaft, placing the screw which impels the rollers upon one shaft, the wheel carrying the knives upon the other, and connecting the two by toothed wheels of varying diameters, and capable of change at pleasure; this produces a variable rate between the velocity of the rollers and the revolutions of the knife wheel, and the hay or straw is cut into lengths proportionate to such variation.

By means of a plate called the presser, the material is secured close together, and this plate in the patent engine instead of being fixed to the support of the upper roller, has a motion round the axis of it, and thus, if the feed is thin, the presser follows down, or if thick, rises up so that at all times the proper pressure is applied. The fault of previous engines being, that the proper pressure was insured only when the feed was uniform; another advantage is also thus gained, inasmuch as no loss of power takes place, for whereas in the old form of engine, a feed that was too thick, was pressed also too lightly and a feed too thin not pressed at all, and the work thereby deteriorated; in this engine, a slight pressure is uniformly given, which, while it answers the purpose, opposes but little resistance to the passage of the material. The parts being strongly constructed, a considerable velocity may be given to the wheel carrying the knives, and from 300 to 350 cuts may be made per minute, through an area of 30 to 50 square inches; 12 cwt. of hay may be cut into half-inch lengths, per hour, with the power of two horses, and the chaff is so uniform as to require no subsequent sifting.

SECTION XI.

THE TURNIP CUTTER.

This machine, as most of our readers will readily remember, must be an instrument of modern invention, for it is but a few years since the cultivation of the turnip has become general in this country.

Although there are several kinds of turnip cutters, the principles upon which they are constructed do not embrace much variety: setting aside the simple application of the knife with a lever handle, the others may be divided into two classes; first, those which have their knives placed on a disc; and secondly, those with their cutting edges arranged on a cylinder.

As the object to be effected is simple, and involves little mechanical contrivance, a short description will suffice. I subjoin a sketch of that one which appears to be the most convenient of any with which I am acquainted; the disc is attached to the side of a barrow, which serves as a hopper; the knife is nearly the length of the radius, and when required to cut the turnip in slices is alone used; if it be necessary to cut small slices for sheep, cross knives are by a simple contrivance adjusted to dissect the slice; and in this case the barrow is useful, as it is easily moved from trough to trough, into which the small slices may be made to fall.

RANSOME'S BARROW TURNIP CUTTER.

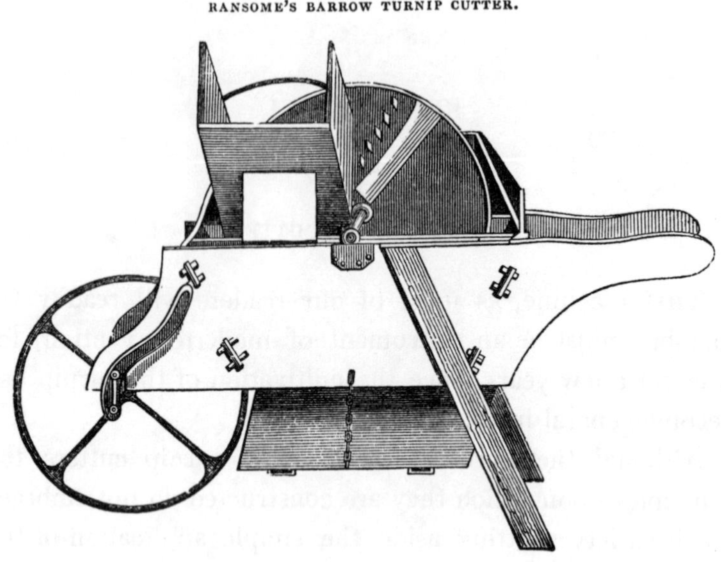

The illustration I adopt for the cylindrical cutter is one which, though of recent invention, yet is now so generally known as to need no further description than is afforded by the wood cut given below. It is intended to cut into small slices for sheep, and is generally acknowledged to be the best implement for the purpose that is at present in use. Our farming readers will not fail to recognise in it *Gardner's Patent Turnip Cutter*.

The following cut shows the Turnip Cutter as it is in use.

GARDNER'S PATENT TURNIP CUTTER.

The following cut shows the same Turnip Cutter as the above, with the hopper removed. Here the peculiarity of the arrangement of the knives is seen. Each knife having cutting edges at right angles, is placed above another till they approach each other at the centre. The cylinder is furnished with two sets of these knives, so that at each revolution thirty cuts are made.

INTERIOR OF GARDNER'S PATENT TURNIP CUTTER.

We here subjoin a sketch of an ingenious adaptation of the disc turnip cutter to the turnip cart. The disc is put in motion by a face-wheel fixed upon the nave of the cart wheel, which, as it revolves, communicates by means of cog-wheels with the axis of the cutting plate. It offers a very convenient mode of feeding sheep on pastures or lawns, and was introduced about the year 1834, by Arthur Biddell, farmer, of Playford, the inventor of the well-known scarifier, which bears his name.

TURNIP CART, WITH CUTTING APPARATUS ATTACHED.

SECTION XII.

MILLS, CRUSHERS, ETC.

These machines, in some rude form or other, have been employed in rural affairs from a very remote period. And even the knowledge of the advantages of bruising, or breaking the food of live stock, is not a modern discovery; for Samuel Hartlib, in his "Legacie," many years since, mentions the advantages of breaking the corn given to horses and cattle.

The term mill seems to have signified originally an engine for grinding corn, but it is now used in a general sense to denote a great variety of machines, whose action depends chiefly on circular motion.

The machinery by which it is necessary to accomplish the ultimate objects of the mill must obviously vary almost indefinitely. Many works on this subject have been published, as well as separate accounts of particular structures. See Brewster's edition of *Ferguson's Lectures; Gray's experienced Millwright; Buchanan on Millwork*, by Tredgold; *Banks on Mills; The Repository of Arts, &c.* A catalogue of the principal works on the subject of mills is given in *Gregory's Mechanics*, vol. 2.—but in this place I shall confine my attention only to those designed for the use of the farmer. In the first place, to such as are used for preparing grain for cattle; and in the second to those intended for the more important object of reducing grain to flour as food for man.

The *Kibbling Mill* is well worthy of notice. It is composed of a small iron cylinder, usually about eight or nine inches wide, and six inches in diameter, tapering slightly to one end, and fluted on the inside. Within this a barrel of the same form, but a size smaller, and fluted spirally on the outside, revolves by the turning of a spindle on which it is fixed. The meal is rendered finer or coarser in proportion as the working barrel is set nearer to or farther from the small end. This mill is made entirely of iron or steel, and is usually attached to a post. It is provided with a hopper, and is worked by a crank fixed at one end of the spindle, while a fly-wheel revolves at the other. It is used for beans, peas, and other pulse, for malt and various kinds of grain, and is a very useful and ingenious contrivance, but requires care in its adjustment and general management.

Referring to Fruth's patent, 1768, it would appear that the credit of the invention of this useful mill belongs to him; but it has been extensively manufactured by Zachariah Parkes of Birmingham, and, from the excellence of its manufacture, they have become so associated with his name as to be generally known as Parkes' Mills.

Bean Mill. A mill for grinding beans, constructed by Seaman and Bryant, of Melton, in Suffolk, is a simple and effective implement. It is placed on a wooden stand, with crank, fly-wheel, and hopper; and consists of a coarsely-fluted steel barrel, working against a cast-iron front cutting plate; the latter being set at a proper distance from the barrel by means of a screw. It is used chiefly for beans and peas, but may be employed for grinding malt, by exchanging the barrel and cutting plate for a pair of rollers.

SEAMMEN'S BEAN MILL.

The *Norfolk Crusher* is similar in appearance to the foregoing, and is worked by two rollers of equal dimensions, each being flanched at one end, and reversed so as to prevent the grain from falling off at the side. The rollers are perfectly smooth, and consequently, as its name implies, it crushes the grain instead of cutting it.

RANSOME'S SUFFOLK CRUSHER.

The *Suffolk Crusher* is simply a variety of the above, and differs from it in having its hind roller finely grooved, and of half the dimensions of the front one ; this has no flanche, but works within the flanches of the front roller, which are attached at both ends. To render these mills effective for crushing oats, the rollers should be left slightly rough as they come from the lathe, to draw in the kernels, as the latter are apt to start back at the moment of entering between the rollers, if they are polished.

The *Spiral Mill*—An improvement, however, on all the previously described mills, and combining in a great degree the peculiar advantages of each, is now obtained, by spirally grooving two cylinders, which, by means of toothed wheels of different dimensions are made to revolve towards each other at unequal speeds, by this means, the tooth or edge formed by each groove in one cylinder passes in a diagonal direction the grooves cut in the other, and the effect produced is similar to that of a series of shears. Machines on this construction will effectually reduce either grain or pulse with less expenditure of labour than any others I have seen.

THE SPIRAL MILL.

Whether mills for the reduction of grain into flour may be regarded as *Agricultural* machines, is a question that will admit of controversy; but as flour mills are everywhere required in the rural districts; and as they must be more or less objects of interest, not only to the farmer, who raises the corn, but to every individual who consumes it, I shall not hesitate to proceed at once to the consideration and description of such as have from time to time been introduced to the public.

It is, however, not my intention to occupy these pages with an elaborate description of the machinery used in the general mill establishments, which has already been done very ably in the works before alluded to, and especially in those periodical magazines that are devoted to the publication of newly patented inventions. Indeed, it would be impracticable to give, within my limited space, any detailed description of those numerous contrivances which modern art has successively introduced into the operations of the public miller on the great scale. Without overlooking the progress of invention as applied to mealing generally, I shall chiefly apply myself to a concise description of those portable machines, which have of late years been so much in request, by agriculturists as well as private individuals, for the reduction of grain to a fit state to be used as food.

Previously, however, to entering upon these descriptions, it may not be improper nor uninteresting, to take a rapid historical retrospect of the mills of olden time: from which it will be seen that the two principal forms of mills now in use, namely, that of cylinders and the frustra of cones are of very ancient date.

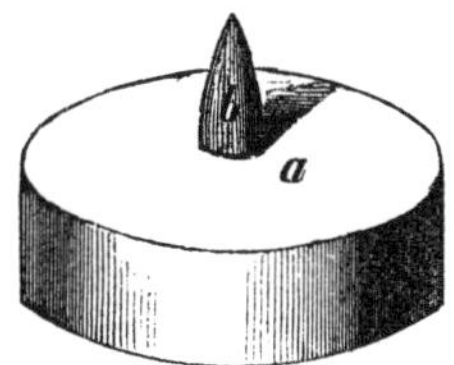

The domestic mills used by the ancient Egyptians, Hebrews, and other nations of antiquity, generally consisted of a circular flat stone (*a*), having an upright pin (*b*) in its centre; this pin, which constituted the axis, passed through a hole (*c*), in an upper stone (*d*), which was provided with a handle (*e*), for causing it to revolve over the lower stone. The upper side of the top stone was hollowed out into a shallow basin (*f*), to contain the corn, and to permit it to fall down the hole (*c*), around the axis (*b*). The annexed figure shows both stones together in their working position. In this beautifully simple apparatus, constructed several thousands of years ago, we behold the germ and the rudiments of our boasted machines of the present day! It was constructed on a principle of action the most efficacious; requiring nothing but accurate workmanship and the means of regulation, to make it produce uniformly the finest results. It is indeed a remarkable fact that the united skill of successive machinists through so many ages, has done little more than adjust and regulate the original contrivance; not from the lack of talent, but simply because the ancient principle of construction was the most perfect that the mind of man could conceive, or will probably yet discover.

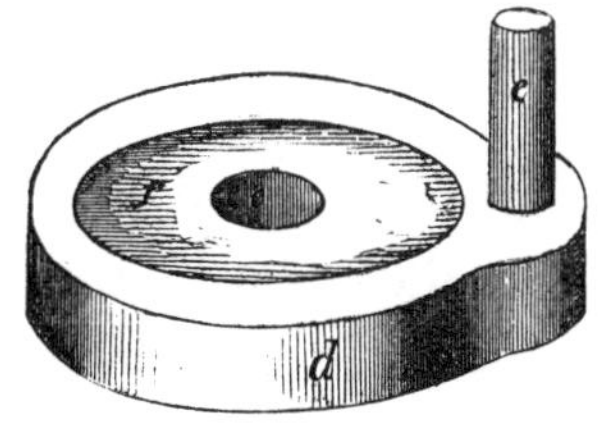

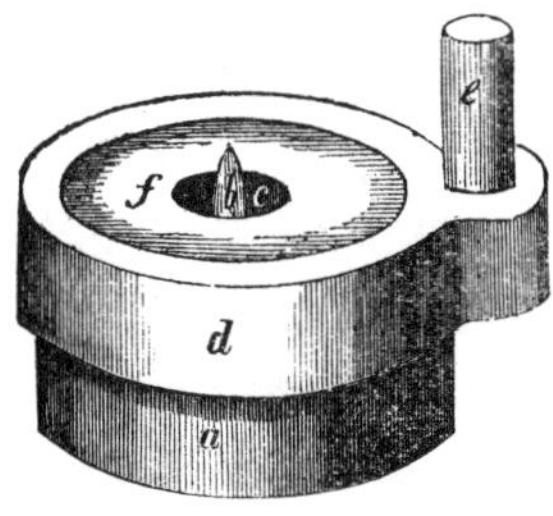

The mill just described, though used as a single-handed mill, and to be held in the lap, there is much evidence to show was occasionally made of larger dimensions; and was sometimes worked by several persons at once, causing it to rotate through the medium of levers, in the manner of a capstan: asses and other cattle were also, in some places, employed as the propelling power. To these the Romans added the force of water to the working of mills. A writer of the time of Cicero thus poetically alludes to this important epoch of the mealing art:

"Cease your work ye maidens, ye who laboured in the mill: sleep now, and let the birds sing to the ruddy morning; for Ceres has commanded the water nymph to perform your task; these, obedient to her call, throw themselves upon the wheel, force round the axle-tree, and by these means the heavy mill."

An illustration of the conical form given to ancient mills (before adverted to) is afforded by the interesting discoveries made in the city of Pompeii, which, in A.D. 79, was suddenly buried in the ashes thrown upon it by an irruption of Mount Vesuvius. Upwards of seventeen hundred years have passed away since the occurrence of that event, and now are brought again to light by the removal of the earth from above the long-buried city, the machines and processes used in several handicraft operations. Amongst these is a large and effective hand mill, standing in a baker's shop, the owner of which appears to have been a wealthy man from the magnitude, solidity, and completeness of his apparatus. The mill is about six feet in height, and five feet wide at the base (*d*), in the centre of the latter is a conical projection (*b*), and over this is placed a double hollow conical block (*c*) (*c*), the upper portion constituting the hopper, and the lower portion the revolving grinder, which is put in motion by levers inserted into holes made in an iron band (*d*). In the fixed grinder (*b*) is inserted an iron spindle, which passes through the revolving one at the bottom of the hopper; and through the latter are drilled holes for the passage of the grain.

POMPEII MILL.

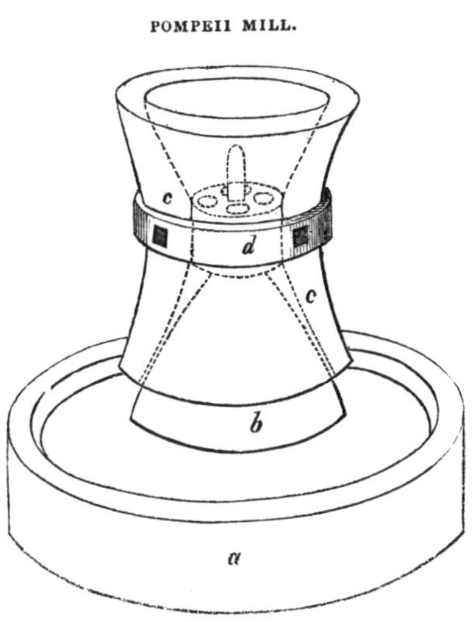

From a variety of circumstances it would appear that the conical form of mills just described did not obtain very generally in Europe, though there is much evidence to show that machines having a resemblance to them are (and have been for many centuries) extensively used in India and other parts of Asia; from whence, indeed, it is highly probable they originated. The ancient mills that we meet with in our own country are generally of the cylindrical form, and of comparatively small dimensions. As civilization advanced the reduction of grain into bread flour became a distinct occupation; that of the public miller. By constant application to his craft, experience and skill were gradually acquired, the power of wind and falling water were advantageously applied, and by degrees the dimensions of his machinery were augmented, until the mill-stones reached the extraordinary size of six, seven, and even eight feet in diameter; in short, every available means which practice or the knowledge of the time developed were put into requisition for economising the process and improving the results. Till at length, by the continual accession of improvement in every department, a modern steam corn mill, with all its self-regulating apparatus for making the varied operations of the vast machine act harmoniously together, under every change of circumstance, presents to the mind an interesting example of human ingenuity applied to an object of universal utility. Although the establishments of the public miller throughout this country have for several centuries past almost superseded the employment of small domestic corn mills; it has been in some measure owing to the non-existence of efficient and durable hand machines that they have not been more extensively used. For otherwise it will be readily supposed that the farmer who has to send, in some cases, many miles to the mill, would avail himself of the means of doing at home that which costs him time of horses and men, as well as other expenses, at an inconvenient distance. It is only within a comparatively recent period, that suitable economic machines for the purpose have been placed within his reach. With the view of obtaining the fullest information on this subject, the writer has attentively examined the several specifications of patents for mills that have been deposited in the enrolment office in Chancery, and although much of the information thus obtained relates to contrivances not needful to particularize in detail, still he conceives that a *very concise* abstract of such of these as immediately bear upon the present subject will be found of utility to all those who have occasion to employ mills, or who take an interest in the progress of mechanical invention, and that presented on this form the information will be most readily obtained by the reader. A slight attention to the respective claims to invention set up by each succeeding patentee, will in general show pretty clearly the degree of originality that belongs to it; succeeding inventors will also find the abstract an important aid in the prosecution of their plans, as by its help they may perchance save themselves the trouble of re-inventing contrivances that have long been in use; and the consequent disappointment and vexation that attends the subsequent discovery.

ANALYTICAL ABSTRACT OF PATENTS GRANTED FOR MILLS FOR GRINDING AND DRESSING GRAIN AND OTHER SUBSTANCES, CHRONOLOGICALLY ARRANGED.

"*To Isaac Wilkinson, of Wilson House, Cartmell, Lancashire, 'for a new sort of cast metallic rolls, for the crushing, flattening, bruising, or grinding of malt, oats, beans, or any kind of grain; and also for crushing, bruising, or grinding of sugar canes.*" Patent granted 24th January 1752.

This invention is of extreme simplicity; it consists in merely passing the materials above mentioned between two plain metallic rollers. It is a process that has ever since been employed very extensively in crushing malt and oats, and also sugar; under various modifications, wherein the original machine has been improved, especially as respects the giving different velocities to the opposing rollers.

"*To John Milne. 'A machine for dressing the flour of wheat and barley, which will make a more lively and better flour than bolting cloths (which is the common method now used) from the same corn. It will dress all sorts of flour, and divide the sharps from the bran at one operation, and the person that attends it may easily make two sorts, or only one, by moving the partitions that divide the flour, which must be within the box or case in which the machine works; and as flour is an article that loses every time it is stirred, it evidently appears that it dresses with less loss because it does that business at one operation, which, to be done with cloths in the common method now used, requires several operations and several different cloths, and the trouble of changing them, they being obliged to change their cloths for different sorts.'"* Patent granted 10th May 1765.

A cylinder six feet long and 16 inches internal diameter (or any other size which may be preferred) has fixed upon the inside, wire work, or cloths of different degrees of fineness. Through the centre of the cylinder passes a spindle having fixed upon it *brushes either lying parallel to the axis*, or *forming a screw or worm*. The cylinder is fixed in a box, *either in a horizontal or a slanting position*. The brushes are made to revolve, and the cylinder may either be stationary or revolve likewise:

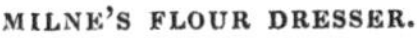
MILNE'S FLOUR DRESSER.

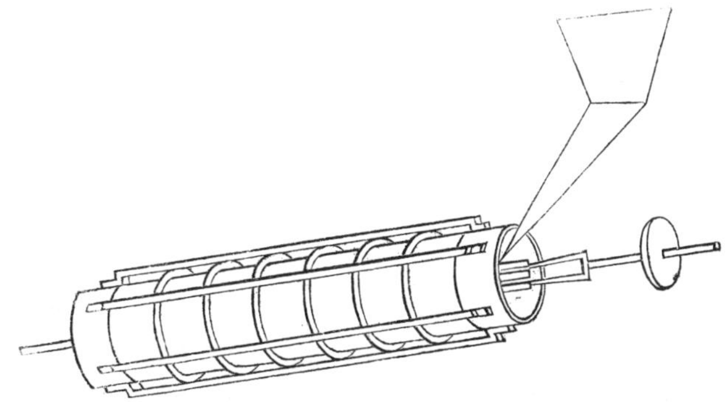

This invention has the conspicuous merit of being the original contrivance of the now almost universally used dressing cylinder. The principle of its construction is so excellent, and its operation so efficacious, as apparently to render any essential improvement impracticable.

To Richard Hayne, of Ashbourne, Derbyshire, 'for a machine or mill so contracted and effectual that it may be set up and conveniently worked in any small room, and used as well for the grinding of wheat, corn, and other grain, as in preparing of utensils and materials used in divers manufactories and businesses. Patent granted 24th December 1767.

If the foregoing title of Hayne's patent be calculated to excite the reader's curiosity to learn the construction of a mill which, by its *contraction,* is to perform such extensive operations; his surprise will be great upon being informed that the inrolled specification describes no mill at all! and that the ingenius patentee was contented with claiming the employment of the motive power of *a smoke jack, assisted by manual force!!*—for the grinding of wheat, and performing all and singular the divers operations set forth.

" *To Samuel Fruth and Samson Fruth, both of Birmingham, in the County of Warwick, merchants, 'for certain hand corn mills for grinding wheat, in private families, in a more easy and expeditious manner than hath been done by any other mill or machine, calculated for private families, heretofore invented.'* " Patent granted 6th October, 1768.

FRUTH'S HAND CORN MILL.

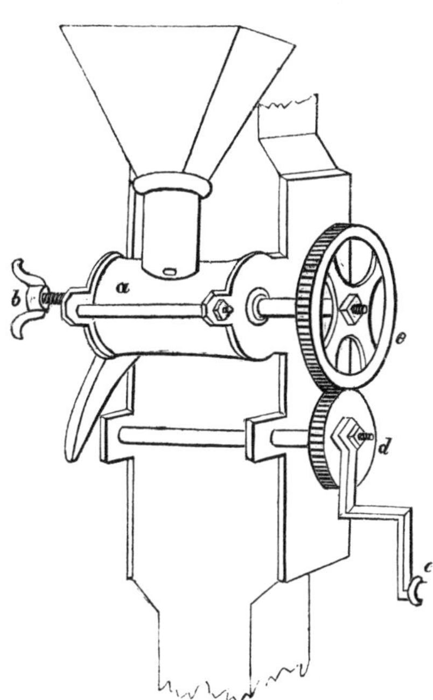

This invention appears to be the earliest patent granted for improvements in metallic domestic corn mills. The above sketch is copied from the specification, which, like many of such documents at that period of the time, seems to have been drawn up with the intention of affording as little practical information as possible as to the construction of the essential parts of the machine. The box (*a*) is stated to contain a roller, which roller, as well as the box, is made a little tapered. The box and roller are both to be full of fine teeth, and both are to be made of *refined iron*, and *case hardened;* (*b*) is the regulating screw; (*c*) the winch for giving motion, through the medium of a pinion (*d*); and a wheel (*e*) to the grinding roller. The slowness of motion is remarkable, being only about half a revolution to one of the handle; the general practice now being to increase rather than to diminish the velocity! The object was probably to gain sufficient power to turn a mill of the size.

"*To Walter Taylor 'for an invention of a great improvement in the construction of machines for grinding grain of all kinds, and also starch for hair powder as well as all other matters where stones are now used, and also for coaking or bushing and greasing of shivers and pulleys of all kinds.'*" Patent granted 30th October, 1768.

This patentee, like the previous one, is very cautious of explaining himself. The substance of the information given in that portion of his specification which relates to grinding is as follows.

"The machines for grinding are of two kinds:—

1st. A machine of cast iron, made nearly like mill stones, with holes or impressions therein to cut and grind the corn, and with grooves or furrows to admit the flour to pass while grinding.

2d, Conical mills of cast iron, with grooves cast on the external and internal surfaces; or with steel cutters fixed in grooves cut on said cover.

3d. Making those mills called steel mills, of cast iron."

"*To Richard Dearman, of Birmingham, in the County of Warwick, Iron Master, 'for making mills for grinding malt, and various other articles, in the same manner as those articles are ground in what are commonly called steel mills.'*"

Patent granted 22d March, 1779. It does not appear that Richard Dearman has much claim to originality in the matter of his patented invention; all he appears to have attempted was to make mills of *cast iron*, instead of steel. There were five or six patents before this for making mills of cast iron, (including those for bark, &c.), yet no one, perhaps, to give the unfortunate speculators the information.

"*To Charles Smith, of Manchester, for a machine for bolting or dressing of flour and meal.*'" Patent granted 4th Dec. 1781.

SMITH'S FLOUR DRESSER.

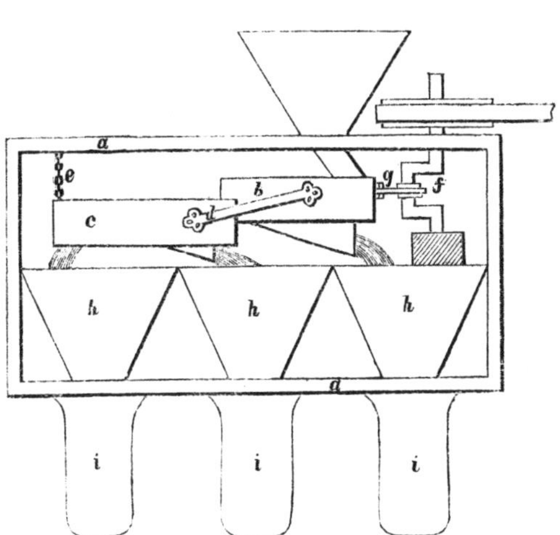

Smith's machine consisted of a close rectangular box (*a*), containing a series of open sieves (*b c*), having any desired number of compartments, connected together by straps (*d*)

at the sides, and suspended at the end by chains (*e*). Agitation was given to the sieves by a crank and rod (*f*) (*g*), the products of flour being delivered into hoppers (*h h h*) which discharged it into bags (*i i i*), underneath the box.

"*To Major Pratt, of Running Waters, in the parish of Pittington Halgarth, in the County Palatine of Durham, lime burner, 'for a method of manufacturing or making a certain composition stone, which will be equally applicable in grinding all species of corn, and other articles, as the mill stone at present used.*" Patent granted 11th March, 1796.

The principle of this invention consists in mixing silicious and argillaceous earths, under certain circumstances, and by a due application of heat, converting them into a substance capable of being formed into mill stones or substitutes for mill stones. The proportion of the above mentioned earths may be considerably varied. As clay and sand alone are highly refractory, the patentee mixes with them about one-seventh part of calcareous earth, previous to their being exposed to the fire. But this proportion is also subject to variation. The patentee also anticipates that other bodies may be discovered that will assist the fusion of the clay and silex, and all such combinations he considers to be within the meaning and scope of his patent. His claim to originality being the conversion of clay and silex, by the action of fire, and with the addition of a third body, into a substance such as above stated and for the before mentioned use.

"*To Robert Ferryman, of Hammersmith, in the County of Middlesex, Clerk, 'for a machine for blanching, grinding and dressing of corn.'*" Patent granted January 24th, 1797.

The contrivances specified under this patent exhibit much ingenuity misapplied. The principal mechanism of a novel character, is for the purpose of "taking off the outer skin or corolla of the wheat." A conical wooden roller is cut with oblique grooves, over which are nailed strips of leather; against the surface of the latter is brought into action a strong cloth, which passes over a small cylindrical roller on each side of the conical roller, and is strained tightly by a suspended weight. The corn descends from a hopper, and is, by the aid of a fixed brush and a leather rubber, spread out over the cone, and by the oblique grooves on the latter is carried round within the cloth, and against the leather rubber which rubs off the coat, and then the corn falls down before a fan, which blows away the chaff, and delivers the blanched wheat into the eye of the mill.

This mill consists of a solid conical stone, with furrows of the usual kind; and outside there is fitted a hollow cone, formed of two stones nicely fitted to together. By these the corn is ground, and the meal is discharged through the centre of a circular brush, working in a circular sieve. The bran is thrown out of the sieve by the centrifugal action, and the flour descends through it into a drawer below.

"*To William Hunt, and Wastel Cliffe, of the Brades, in the County of Stafford, Steel Manufacturers, 'for a method of grinding corn, malt, and other grain, with steel or iron hardened plates.'*" Patent granted, 2nd August, 1799.

The patentee's directions are, to take circular plates of steel, or hardened iron, and to roughen them considerably with lines, grooves, or furrows, disposed according to the nature of the work required. To the upper grinder are to be screwed certain metallic projections, which, he says, are to rub the case, and clear the grinders of the flour, which, he says, is effected in stone mills by the inequalities of the surfaces of the stones. The drawings given in the specifications, and the written description are both defective.

"*To Thomas Wright, of Mark-lane, in the City of London, Broker, 'for an improved method of making hand-stone corn mills, for the purpose of grinding wheat and other grain into flour.'*" Patent dated 25th April, 1801.

WRIGHT'S HAND-STONE CORN MILL.

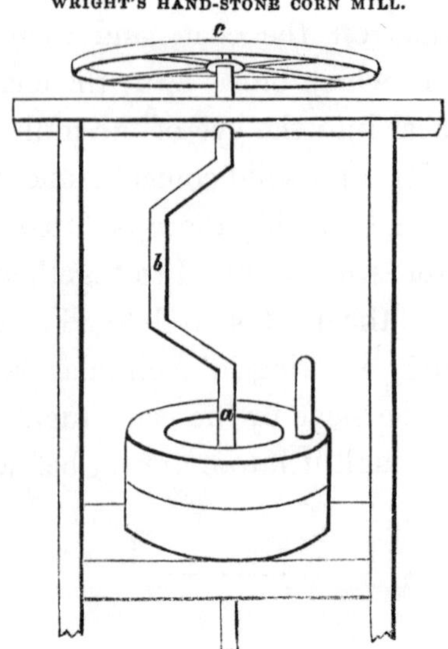

The annexed sketch is a representation of the " improvement" specified. The stones are said to be from 12 to 15 inches in diameter, to be finely grooved for about 3 inches from the centre, and pecked over the remainder of their surfaces. The runner stone is supported upon a spindle, (*a*) resting on a bridge, and is turned by the crank, (*b*) and at the upper end of the spindle is a fly-wheel (*c*). Although mills were frequently mounted in this manner long before the patentee invented it, the form is not without its advantages for some purposes, especially for the uses of the laboratory. The inventor has been still more unfortunate, in an historical point of view, in the concluding part of his patent; wherein he directs that instead of the crank, a handle like that shown by dots may be inserted in the upper stone, to turn it by. This, as will be noticed upon reference to p. 205, was in use about 4000 years ago, and has been ever since.

" *To Charles Williams, of Gravel-lane, Southwark, Millwright, 'for a machine for grinding or cutting malt, splitting beans, and other kind of grain, and various other articles.'* " Patent granted 2nd August, 1810.

This machine is composed of a cylindrical or conical roller, made of cast-iron, or any other metal, with grooves cut in it, in an oblique or parallel direction; this roller acts against a series of knives, screwed together so as to form the same curve as the roller, and so that they may be taken out and sharpened at pleasure.

The annexed sketch represents a side elevation of his mill

for grinding malt, designed for the reduction of large quantities.

WILLIAMS' MALT GRINDER.

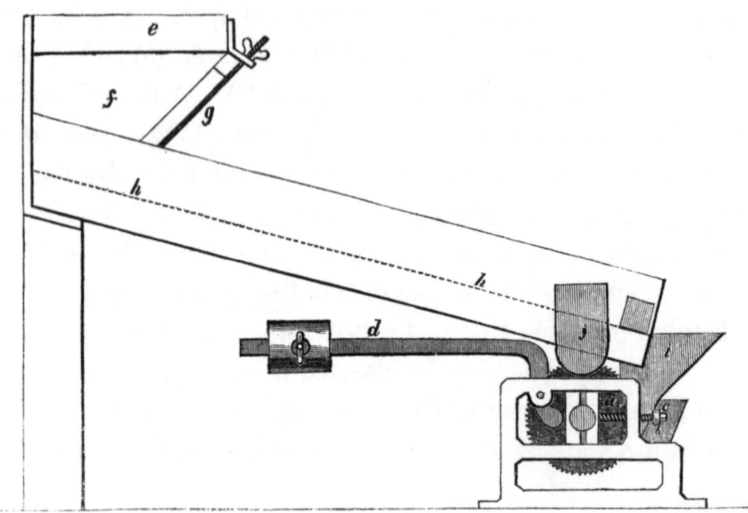

At *a* is the grinding roller, the axis of which is supposed to be connected with the prime mover. In a frame at (*b*) are a series of five or six knives, fixed in a direction parallel with the roller (*a*.) At (*c*) is shown one of two adjusting screws, to keep the roller at a proper distance from the knives: (*d*) represents one of two weighted levers, the short arms of which are made to press against the brasses of the roller, to keep the latter up to its work, but at the same time to permit any hard substances to overcome the resistance of the lever, and to be ejected from the mill. At (*e*) is a coarse sieve, to permit no substances larger than the malt to pass through into the hopper (*f*); whence the grain, regulated by a sluice (*g*), passes over an inclined screen (*hh*), that allows the dust and small dirt to escape, and deliver the cleansed material into a feeding hopper (*i*,) the latter delivering it between the grinding surfaces of the mill: (*j*) represents one of two beaters, which, by means of projections on each side of the roller (*a*,) causes the inclined screen (*h h*) to be continually agitated during the process.

" *To George Smart, of the Ordnance Wharf, Westminster Bridge, Surrey, Timber Merchant, 'for certain improvements in machinery for grinding corn and various other articles.'* " Patent granted 1st April, 1814.

This invention is designed to supersede the use of mills for the purposes mentioned, and to substitute for the grinding process that of breaking and crushing. No drawing accompanies the specification; and the description, as well as the claims to invention, are so general and indefinite, as to be quite inadequate to enable any mechanic to construct the apparatus. The breaking and crushing is to be "performed by the application of rubbers or crushers, resting on their fulcrums," and to be "pressed against the revolving body by means of levers, weights, or springs."

" *To Archibald Kenrick, of West Bromwich, in the County of Stafford, Iron Founder 'for certain improvements in the mills used for the grinding of coffee, malt, and other articles.'* " Patent granted 23d May, 1815.

Kenrick's invention related more to the external fittings-up of the mill, than to any internal improvements of the working parts; the latter consisting, as before, of eccentric frustra of cones, with vertical axes for the portable box mills, and horizontal axes for the post mills. It is, however, but justice to say that Kenrick's improvements in *manufacturing* all the parts were so judicious and complete as to render these little domestic machines much more durable and convenient than they were previously.

" *To Francis Devereux, of London, 'for certain improvements in the machine for grinding wheat and other articles, commonly known by the name of the French military mill.'* " Patent granted 8th Jan. 1824.

In this invention are employed metallic plates with radiating grooves, turning in a vertical plane. The fixed plate is bolted to the end of the box framing, in which the mill is enclosed. The revolving plate is fixed to an axis, one end of which revolves in a bush in the boss of the fixed plate: and the other end in a bush in a bracket, which is bolted to the lugs of the fixed plate. The grain enters the centre of the fixed plate, through a channel at the back of it; and the meal as it is thrown off at the periphery, falls into a hopper beneath the grinders, and passes out of the mill through a hole at the bottom of the box.

As the manner of regulating this mill was the point on which the patentee rested his claims, it may be proper to describe it, with reference to the annexed section of the machine.

DEVEREUX'S CORN MILL.

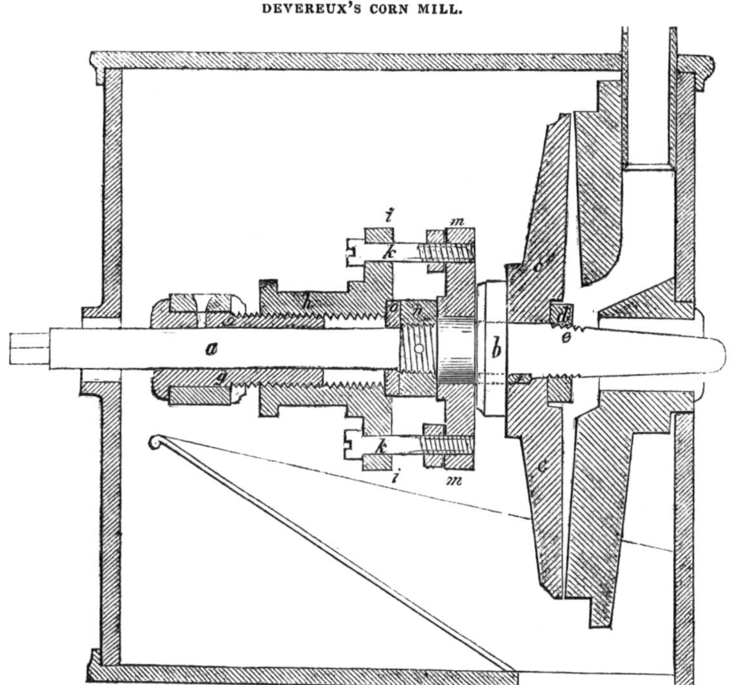

At *a*, is the shaft, having a shoulder at (*b*,) against which the running grinder, (*c*) is firmly pressed by the nut (*d*,) which works on a screw, (*e*) out on the shaft; the grinder being prevented from turning round by the pin (*f*). The bush, (*g*) in the bracket which supports one end of the shaft, projects beyond the bracket, and has on the outer end an external screw. The regulating apparatus consists of a tube, (*h*) having an internal thread, fitting the screw on the bush, (*g*) and arms, (*i i*) which are connected by bolts (*k k*) to a collar, (*m m*) placed upon an enlarged part of the shaft. The collar, (*m*) is retained in its place, at the back of the collar, (*l*) by the nut, (*n*) screwed and pinned to the shaft, leaving sufficient play between it and the collar, (*m*) to allow the shaft to revolve freely within the collar, and a washer (*o*) is placed between the arms, (*i i*) of the regulator, and the nut (*n*). The regulator is prevented from turning by a click or pall working in notches on the regulator, so that the collar remains motionless whilst the shaft revolves; but when it is required to regulate the distance between the plates, the pall is lifted out of the notch, and the regulator is then screwed forward or backward, which by means of the collar, (*m*) moves the shaft endways in a corresponding direction.

"*To James Ayton, of Trowse, Millgate, Norfolk, 'for an improvement to be applied to Bolting Mills, for the purpose of facilitating and improving the dressing of flour, and other substances.'*" Patent granted 19th February, 1825.

These improvements are intended to produce a more powerful vibration of the bolting cloth, and to render its action upon the beater more equable. For this purpose a series of springs, capable of adjustment, and proceeding from a common centre like the radii of a wheel, receive the the loops at the tail of the bolting cloth, and thus give at the same time that degree of tension and elasticity which is necessary to produce a uniform and powerful vibration of the bolting cloth. These springs are composed of steel arms riveted to a ring of metal, which is secured to an internal ring by means of two conical pointed screws, and this ring again is secured to a collar or boss on the shaft by two other conical pointed screws standing at right angles to the former; the four screws by their combined action forming a kind of universal joint upon which the steel arms are easily adjusted to the required position. The outer extremities of the arms are formed into broad hooks to receive the loops of the bolting cloth, for which purpose they are nicely rounded and smoothed.

"*To John Smith, of Bradford, 'for certain improvements in Machinery for dressing flour.'*" Patent granted 4th June, 1829.

These improvements consist *first* in using iron ribs for forming the frame of the dressing cylinder; *secondly* in a method of fixing the wire-work thereto, and *lastly* in using an external brush to cleanse it.

The ribs are formed of two semi-circular pieces of iron jointed together by screws and nuts, and kept firm by two bars at opposite sides passing through transverse perforations made in each rib; through the latter, in a contrary direction, are also formed numerous other perforations, for the admission of screws, which pass through the wire-work, and the holes prepared for them, and are then screwed firmly by finger nuts, whilst the heads of the screws fit into a groove or channel formed along the inner circumference. A brush is placed above the cylinder and both are caused to revolve, though with different degrees of swiftness, by means of gear properly arranged for the purpose. The axis of the former turns in a two armed lever, which latter is caused to move easily on a pivot at certain intervals, being acted upon by a second lever attached to part of the gear in order to raise it when the connecting bars in passing would otherwise come in contact with it and prevent its working. An internal brush is also used as in the usual machines. (Repertory of Patent Inventions.)

"*To John M'Curdy, of London, 'for certain improvements in the method of constructing mills, and mill stones for grinding.'*" Patent granted 2d November, 1829.

It will be a sufficient description of the Patentee's invention to state the claims to originality which he makes in his specification; these are,

First. Agitating the boulter (which is situated beneath the mill stones) by means of a projecting pin on the shaft to which the upper stone or runner is attached, which pin comes in contact with a lever, to which the boulter is suspended.

Second. Supporting the vertical shaft in a socket fixed on a bridge hinged at one end, and supported by a screw at the other; whereby the distance between the stones can be regulated at pleasure.

Third. Cutting in the upper surface a number of grooves about a quarter of an inch deep, and proceeding from the centre in a spiral direction to the circumference.

Fourth. Employing a composition for repairing and making mill stones, which consists of a quantity of the French grit stone granulated, mixed with five times the quantity pounded, and an equal quantity of alum; these are to be boiled together, and poured into the holes to be filled up; or into appropriate moulds if used for making new artificial stones.

The patentee also claims, amongst other things, a mill with the stones placed vertically. A mode of construction which, however, possessed no novelty; the Society for the Encouragement of Arts and Commerce having a great many years previous rewarded a gentleman of the name of Russall for a mill of that description.

"*To David Selden, of Liverpool, Merchant, 'for certain improvements in mills for grinding coffee, corn, drugs, paint, and various other materials.'*" Patent granted 11th August, 1831.

SELDEN'S COFFEE MILL.

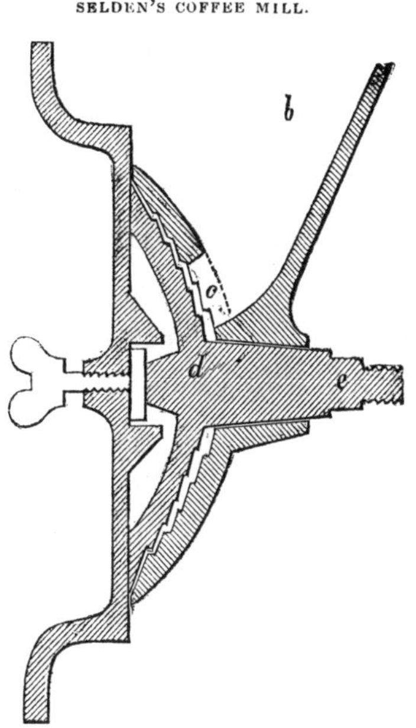

This invention has the merit of extreme simplicity of parts. It consists of two concentric grinders made of metal and with grooves and furrows on their respective surfaces, resembling those of ordinary mill stones. At *a a* is the fixed grinder, having cast to it a hopper *b*, the feed-hole from which is at *c*. At *d* is the revolving grinder, of which *e* is the axis, to one extremity of which is attached the winch, and to the other is applied an adjusting screw.

The edges of the cones are directed to be left plain for about half an inch for the purpose of reducing the substances to an impalpable state.

" *To George Goodlet, residing in Leith, proprietor of the London, Leith, and Edinburgh steam mills, 'for a new method of preparing rough meal from ground wheat or other grain previous to their being dressed for flour; also rough meal from ground barley, malt, or other grain, previous to their being put into the mash-tun for brewing or distilling.'*" Patent granted May 3rd, 1832.

This invention consists in spreading out over a large surface, and to the depth of six inches, more or less according to circumstances, the rough meal of wheat for flour, or of oats for oatmeal. The floor of the drying room should be over a steam engine, or be heated by flues from one, according to the purpose for which it is intended.

The patentee states that the meal or flour is by this means rendered finer, and will produce more abundantly; that new wheat can be ground immediately without any mixture with old, and that flour rather unsound, may by this process be rendered wholesome and nutritious.—*Repertory of Patent Inventions.*

To C. M. Savoye, of Oxford-street, Middlesex, 'for an improvement or improvements in mills or machines for grinding or reducing grain and other substances." Patent granted 15th Dec. 1832.

The grinders in this mill, consist of two concentric rings of metal, having on their contiguous working surfaces a series of teeth cut in such manner, as to present cutting edges, whether moved to the right or left. In the upper part, they project about a quarter of an inch, and diminish in a downward direction, to a plain or smooth surface, so as gradually to reduce the grain to fine flour. The internal ring is fixed to a vertical shaft, which instead of being actuated by a rotary motion, as usual, receives a reciprocating movement through the agency of a crank; causing the internal grinder to turn halfway round, and then back again, continually. This motion causes the feed to be taken very uniformly at every half revolution; the back stroke always clearing the feed-hole. There is a loss of time and also a small waste of power by this singular mode of operating; it has nevertheless the advantage before mentioned, and is, upon the whole, a very respectable contrivance.

"*To Luke Hebert, of London, Civil Engineer, 'for certain improvements in machines or apparatus for, and in the process of, manufacturing bread from grain; and in the application of other products, or another product thereof, to certain useful purposes.'"* Patent granted 24th January, 1833.

The specification of this patent is descriptive of a series of novel machines and processes, designed for operating upon wheat in all its stages of manufacture, until it is ultimately converted into the state of bread; bread, however, in which the "uncertain process" of fermentation is not resorted to, but one whereby the spongy texture of ordinary loaves, called "light" is produced by the introduction of carbonic acid gas, into a close vessel during the process of kneading. The baking is likewise effected by the heat of steam circulating around the ovens. To give any description of the various apparatus specified does not fall within the limits of this work, we shall therefore confine our attention to Hebert's mill; which, whatever may be said of its merits, has at least novelty to recommend it to the notice of the reader. It will not be inappropriate to the subject of this essay to state the inventor's reasons for the introduction of his new machine. Writing in 1834 he says—

"In grinding wheat it has ever been the endeavour of millers to separate all the flour from the husk without pressing it so hard as to "kill it," and without deteriorating its colour by making many minute "greys." This they have not been able to effect in a convenient or profitable manner with the mills constructed on the usual plan, nor by any form of construction that has hitherto appeared. The reason is obvious; if the stones be brought so close together as is necessary to remove the firmly adhering portions of the flour from the husk, the whole of it will be, in a great degree, "killed" and discoloured by the violent rubbing necessary to clean the bran; on the other hand, if the stones are kept further apart, so as to "grind high," much of the flour will be left in the offals and bran.

'With a view of meeting these difficulties, some millers have ground their wheat at two distinct operations; they have, in the first place, set their stones wider apart, or 'ground high;" and then, after collecting the meal, and separating the fine flour from it, have passed the remainder a second time through the stones, setting them closer than before, or "grinding low." Thus have they removed the whole of the flour from the husk, and preserved the good quality of a part of it; but the waste, and loss of time and power in conveying the meal from one place to another, occasioned by these several operations; together with the difficulty of separating the flour from it in the unfinished state by the ordinary dressing machine, have been found to neutralize the advantages otherwise resulting from this mode of proceeding.

"In consequence of the great size and weight of the stones usually employed, the buildings and fittings up of the mill are necessarily very heavy and expensive; and, owing to the several processes of grinding, cooling, "ageing" and dressing the meal, and clearing up the offals, being conducted in situations remote from each other, a waste of flour, together with much unnecessary manual labour, and waste of mechanical power, are incurred.

"These disadvantages, which are inseparable from the old system, are completely obviated by the 'Patent Portable Progressive Corn Mill,' from the following causes:

"Instead of employing only a single pair of stones of great weight and diameter, a progressive mill consists of two pairs of stones, of smaller diameter; with a flour dresser between them, into which the meal from the top pair of stones freely descends. Two-thirds or three-fourths of very superior strong flour is thus at once produced, while the unfinished portion falls into the eye of the second pair of stones underneath. This second pair are set closer together than the first, to complete the softening of the remainder of the meal, which, in consequence of the bulk of the flour being separated from it, is much more easily operated upon. Underneath this pair of stones is placed a common dressing machine, into which the meal falls as it is ground, where the remaining flour, as well as the different qualities of offal, are separated in the usual way."

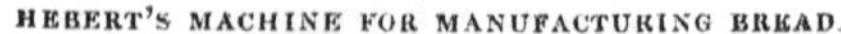
HEBERT'S MACHINE FOR MANUFACTURING BREAD.

At (*a*) is the case of the top pair of stones; (*b*) is the top flour dresser; (*c c*) is the lower pair of stones; one of the shutters of the case being removed to show this part of the machine. At (*d*) is the dressing machine, wherein the work is completed.

" *To Thomas Don, of Lower James Street, Westminster, Civil Engineer, 'for certain improvements in mills; particulars of title to be inserted.* Patent granted 7th March, 1833.

This invention consisted in the employment of a vertical mill stone, turning on its edge, and bearing against a segmented curve of another stone of rather greater radius.

" *To Thomas Sharp and Robert Roberts, of Manchester, Engineers, 'for certain improvements in machinery for grinding corn and other materials.'" Communicated by a Foreigner.* Patent granted 1st Jan. 1834.

This patent originated in an ingenious suggestion on the part of a foreigner, to increase the triturating effect of ordinary grinders, whether of stone or metal, by causing them to rotate *eccentrically* to each other; instead of the usual concentricity of motion. The attempt to introduce this presumed improvement has been ably supported by the celebrated engineers who became the patentees for this country. Superior workmanship, skilfully applied, has left nothing undone which should tend to the carrying out the notions of the inventor. Although I am disposed to doubt the superiority of the principle over that more usually employed, as the machine is one of so novel a character, and so ingeniously constructed, further apology for its introduction to the reader will be superfluous.

SHARP AND ROBERTS' CORN MILL.

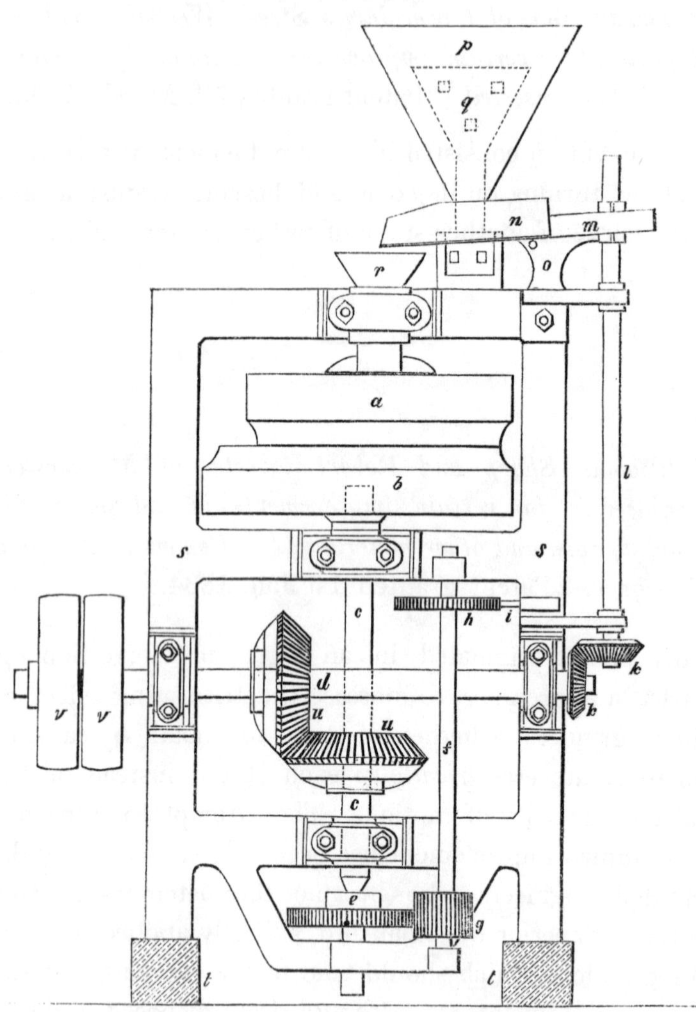

At (*a*) is the upper grinder suspended on a fixed socket, round which the grinder revolves; (*b*) is the under grinder, secured to the main vertical shaft (*c*), which is put into motion by the main horizontal shaft (*d*); at (*e*) is a cup for the foot of the vertical shaft to work in; this cup is fixed to a wheel and screw for raising or lowering the under grinder; (*g h*) pinion and ratchet wheel on the regulating shaft (*f*); (*i*) a spring to work in and secure the ratchet wheel (*h*) in any required position. At (*k k*) are two mitre wheels to work the vertical shaft (*l*) for the purpose of giving vibratory motion to the feeding shoe, by the finger (*m*) striking it periodically; (*n*) is the feeding shoe, with a stud passing through a carrier (*o*) to allow the shoe to vibrate; (*p*) is the hopper, secured to a supporter (*q*); (*r*) is the funnel to carry the grain to the grinders, fixed to the socket of the upper grinder; (*s s*) the framing of the mill, and (*t t*) feet to which the framing is bolted; (*u u*) mitre wheels on the two main shafts of the mill; and (*v v*) a pair of fast and loose pullies to be driven by a strap from a drum, to which the motive power is applied. The grinders are inclosed in a sheet iron casing, to which a spout is attached to convey the flour from the mill to the receiver.

"*To Miles Berry, of Chancery Lane, Middlesex, 'for certain improvements in mills, for grinding wheat and other grain, and which improvements render them applicable to other purposes.* Patent granted 13th September, 1834.

At (*a*) is a mill stone turning vertically on a horizontal axis; (*b*) a curved segment adjusted to the tunnel by mechanism not shown; (*c*) is a case to prevent the escape of flour; (*d*) is a feeding spout delivering the grain at (*e*), which, on its passage to (*f*), becomes ground into meal and ejected from the machine. This invention having been anticipated by Thomas Don's, mentioned above, the patentee, Miles Berry, (who received his communication from abroad) was under the necessity of restricting his claim in his specification to the mechanism employed for adjustment.

BERRY'S CORN MILL.

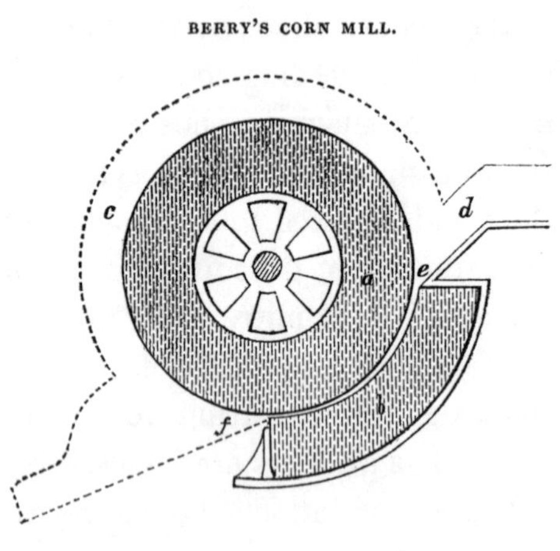

"*To Luke Hebert, of the City of London, Civil Engineer, 'for certain improvements in flour mills.'*" Patent dated 10th August, 1835.

This invention appears to have been the first in which the processes of grinding corn and dressing the meal, were conducted simultaneously in a single machine of great simplicity. The combinations are as follow:—

At *a*, is an axis, which, upon being turned by the winch *b*, (shewn as partly cut off to save space), gives motion to the fly wheel, (*c*) and the mitre wheel (*d*). This wheel actuates another similar wheel, (*e*) at the upper end of a vertical spindle, which passes through the hopper, and carries at its lower extremity the revolving grinder, and some brushes contained in the circular box, (*g*) at the bottom of which is a moveable wire gauze sieve. The revolution of the brushes

HEBERT'S FLOUR MILL.

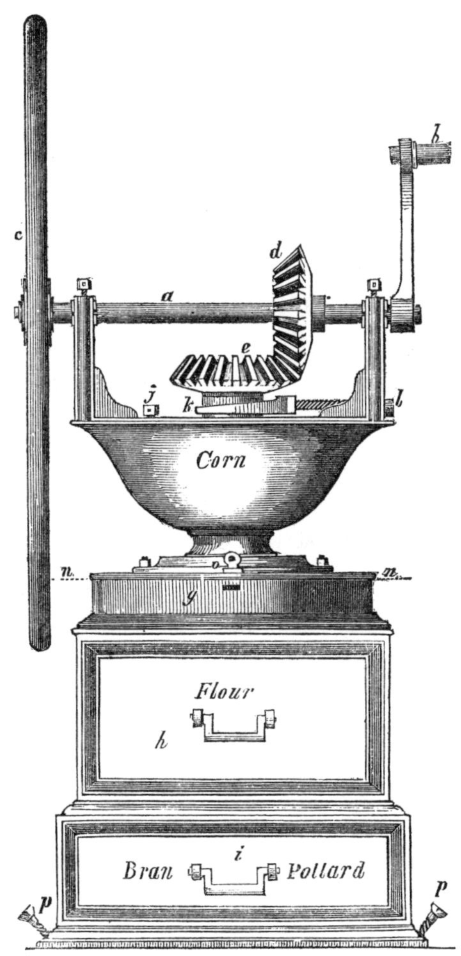

causes the meal as fast as it is projected from the grinders, to be spread over the wire work, and the flour to fall through its meshes into the drawer, (*h*) while the bran and pollard, (which are too large to pass through the wire gauze) are by the centrifugal action of the brushes continually swept spirally outwards, until they are thrown down an aperture at the extreme edge, on to an inclined screen of coarse wire work, which separates the bran from the pollard, and causes each to be deposited in a distinct compartment of the drawer (*i*). To adjust the feed to the effective power of the mill, and the strength of the individual working it there is a screw head at (*j*), by turning which the feed-hole becomes enlarged or contracted For adjusting the quality of the grinding, there is a screw head at (*l*), by turning which forward, a forked wedge, (*k*) draws up the revolving axis, and with it the lower grinding plate, nearer to the fixed one; thereby rendering the grinding finer; consequently, by turning the head (*l*) backward, the reverse effect takes place. The upper part of the machine above the dotted line, (*n n*) is hinged on to the lower, as the lid of a box, and it is fastened down by a turn buckle, at (*o*); by this arrangement, the wire work may be changed or removed, and immediate access be obtained to the working parts.

" *To Luke Hebert, of the city of London, Civil Engineer, 'for certain improvements in mills and machines, for economising and purifying the manufacture of bread.* Patent granted, June 2nd, 1836.

As the improvements alluded to in the above title, have relation solely to the details of the machinery described in Hebert's patent for 1833, it does not fall within the scope of this abstract, to do more than refer the reader to the enrolled document.

" *To William Horsefield, of Swillington mills, near Leeds, in the county of York, corn miller, 'for his invention of certain improvements in the construction of mills for grinding corn.'* " Patent granted March 19th, 1838.

These improvements consist, first, in a better method of fixing and carrying the top or running stone, so that when in full motion, it will be carried round in perfect equilibrium. Secondly, in a superior method of feeding the stones with the corn. The specification exhibits by elaborate drawings, the improvements as applied to mills, driven both by under and over gear. I will endeavour to give some notion of the improvements as applied to a mill of the former class.

The vertical shaft to which the runner stone is attached, is supported by a bridge in the usual way. The head of this shaft is made four-square, and tapering. On this head is fitted a rectangular, cross, the centre hole of which is very accurately formed, so that the four arms shall stand exactly at right angles to the shaft. On the top of the runner stone is bolted a strong iron ring, and to this ring is bolted the four arms of the cross. This contrivance constitutes the first improvement.

To give a clear idea of the second mentioned improvement, some diagrams would be desirable, but I must attempt it without.

Raised three or four feet above the top stone, and supported by a light iron frame, is a small basin or hopper, having a hole in its bottom, from which depends a metal tube, carrying at its lower end a funnel shaped appendage, with its mouth downward, and its neck sliding upon the extremity of the tube. Precisely underneath, and very near to the edge of this inverted funnel, is a small metallic table, or flat disc, of rather larger diameter than the mouth the funnel. This disc is placed (a little elevated) in the centre of a circular dish, having a series of holes at the bottom, and a raised rim at its circumference. The dish and the disc are united; and they are supported upon the arms of the cross before mentioned, consequently they revolve with the top stone; therefore any corn deposited upon the disc, will by the centrifugal force be thrown off, against the rim of the dish, whence falling to the bottom of the dish, it passes through the holes in the latter, into the eye of the top stone. It will now be readily conceived that the object is to continually supply this disc with the proper quantity of corn, according to the varying demand for it. This is effected by a sliding scale; which chiefly consists of a scale beam, to one end of which is a long wire connected with a screw, which regulates the amount of depression of the other end of the beam; whereto is fastened another wire, the lower end of which is connected to the funnel before mentioned, which slides upon the vertical tube. This tube being always kept filled with corn, and the funnel at the bottom of it nearly touching the disc, it is obvious that in proportion to the regulated height of the funnel above the disc, will be the feed to the mill.

"*To Alexander Dean, and Evan Evans, of Birmingham, Millwright, 'for certain improvements in mills for reducing grain and other substances to a pulverized state, and in the apparatus for dressing or bolting pulverized substances.'*" Patent granted 24th September, 1840.

DEAN AND EVANS'S CORN MILL.

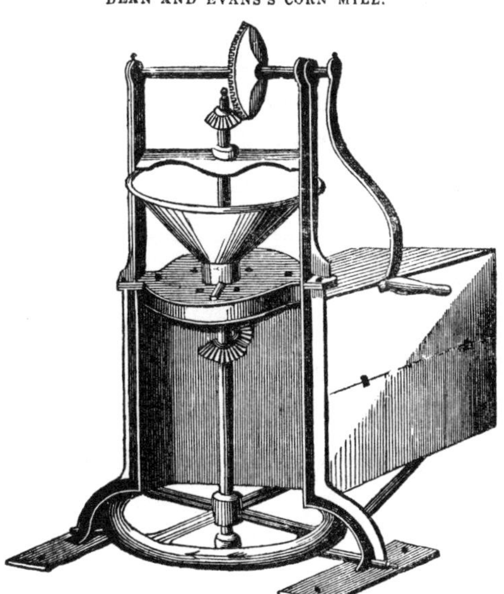

The above wood cut represents the machine published by the patentees. After having described the previous invention of Hebert, this may be very briefly explained. All the upper or *grinding* part of the machine is essentially the same as Hebert's; but the *dressing* is different: Dean and Evans add to the grinding part the dressing apparatus with the spiral screw, or worm brush. These parties claim—

First. The method of constructing grinding surfaces or plates, with a plain or smooth portion, in addition to the usual grinding surfaces of the plate; and they also claim the exclusive use of grinding plates formed of porcelain or earthenware.

Second. The use of a spiral brush, which while it forces the pulverized substances through the interstices of the wire gauze trough in which it works, at the same time forces the substance to be dressed over the upper surface of the wire gauze trough, or forces it through the interstices of the gauze, thus obviating the necessity of placing the bolting or dressing apparatus in an inclined position.

Third. The dressing and bolting apparatus constructed with a spiral brush, whether in combination with a mill or used as a separate apparatus.

"*To George Scott, of Louth, miller, 'for certain improvements in flour mills.'*" Patent granted 23d Sept. 1841.

This invention consists in strengthening the wire-work of a dressing machine against the pressure of the brushes and meal; by placing the circular ribs which support the wire work, in a position oblique to the axis, instead of at right angles to it.

"*To Zachariah Parkes, of Birmingham, 'for an improvement in machines for grinding and dressing wheat, and other substances.'*" Patent granted 13th September, 1841.

Z. Parkes's claim is for the combination of a dressing machine to the old fashioned steel or kibbling mill, mentioned at p. 200. The spout from the mill delivers the meal into the dressing machine, the axis of which is turned by a gut or band from a pulley next to the fly-wheel.

" *To Luke Hebert, of Dover, in the County of Kent, Civil Engineer, 'for certain improvements in machines for grinding and dressing, and sifting grain and other substances.'* " Patent granted 9th Jan., 1843.

The specification of this patent is descriptive of several inventions adapted not only to the grinding and dressing of wheat, but to the comminution, separation, or crushing of barley, oats, rice, malt, sugar, and various other substances.

The woodcut on the next page exhibits a perspective outline of a portable mill for grinding and dressing wheat, denominated by the patentee "Hebert's Patent *Duplex* Flour Mill," the *twofold* operation of grinding and dressing being conducted simultaneously in a single machine, possessing scarcely more complexity of parts than the common dressing machine only.

At (*a*) is a horizontal axis, passing entirely through a cylindrical dresser (*b*), and carrying the fly-wheel and handle. At (*c*) is a hopper, the corn from which enters between the coils of a peculiar Archimedian screw, formed upon the axis, which, by its revolution, necessarily impels the corn forward, and causes it to be delivered between a pair of vertical grinders; the stationary one of which is fixed against the end of the cylinder, and the rotary one is mounted upon the axis of the dressing machine, into which the meal falls direct from the grinders. On to the axis are attached two or more brushes having their bristles worked in an elliptical direction, which causes them, as they revolve, gradually to sweep the meal from one end of the wire gauze of the cylinder to the other; and to separate the various

HEBERT'S GRINDING AND DRESSING MACHINE.

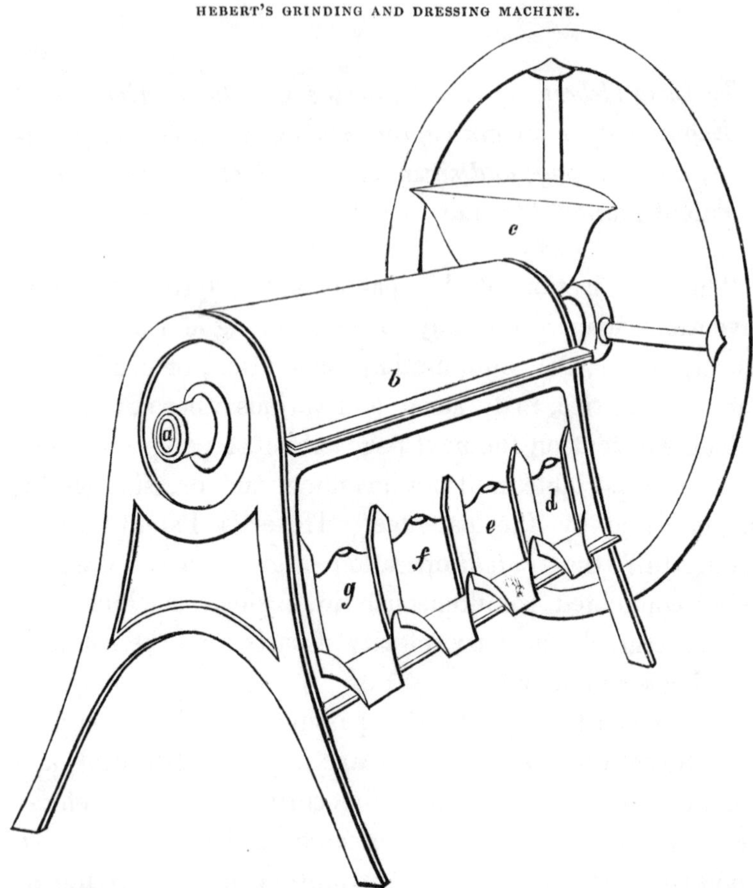

qualities of flour and offal, which deposit themselves in their appropriate bins underneath marked (*d*, *e*, *f*, *g*).

These machines are designed to be worked either by hand or power. The different sizes of wire gauze are so arranged that any one, or all of them, may be instantly changed for others of a different texture; so as to adapt the machine to a great variety of grinding and dressing, or of grinding only.

The grinding plates are susceptible of ready adjustment; they are made of extreme hardness and possess great durability; and they may be changed at any time with the utmost facility.

SECTION XIII.

THE DYNAMOMETER.

The usual dynamometer is an instrument requiring considerable skill to use, so as to arrive at any thing like correct results when it is applied to ascertain the draught of agricultural implements. It is, in fact, a spring-weighing machine, with a hand moving along a graduated arc of a circle, upon which are marked the weights corresponding to the force applied: it does not, however, register the time or space through which the force is in action. An approximation to truth only can at best be made by it, and to accomplish this, a great number of observations must be made, and the average taken, noting the draught of an implement, say a plough or a scarifier. There will often occur a very high indication of resistance from a stone, a root, or some such obstruction, but which resistance is only instantaneous, and if recorded, would interfere with a correct average. Again, we are without any certain knowledge of the resistance to cutting, breaking, and turning over of soil at different velocities; as, for instance, we do not know whether a plough moving at the rate of three miles an hour, meets with more or less, or the same resistance, that it would if moving at two miles per hour, all other circumstances being the same. The dynamometer made by Cottam and Hallen is an attempt at improvement on the common one, inasmuch as it does not indicate the variations so instantaneously, but it is equally deficient in the element of time or space. I understand that R. Clyburn, of Uley, has lately invented one which remedies these defects, and that registers, with precision, not only the power, but also the space through which the power is exerted. I have not seen this instrument, but if it accomplishes this object it will be a much more valuable one than any at present introduced.

SECTION XIV.

STEAM ENGINES.

Having thus far described the implements generally used for the purposes of agriculture, it will not, perhaps, be out of place to draw the attention of the reader to the application of steam power as a means for economically working many of those implements; and if we may judge from the conversation of agricultural gentlemen, we believe there exists a general impression that some of those operations now requiring many horses, may be performed by its aid.

It is not surprising that any one employing horses to draw ploughs should, upon seeing steam engines upon railroads dragging heavy weights at great velocity, at once consider how valuable this power would be to cultivate the soil, and at first sight there would appear to be little difficulty in making this adaptation. Those who are best acquainted with modern progress in every branch of science will be the last to pronounce an opinion as to the extent to which improvements may be carried on or to say that even ploughing fields by a steam engine is chimerical: on the ground, however, that to know the difficulties to be overcome is the first step to success, we would merely state a few facts, that may serve to correct the judgment as well as tend to moderate the expectations some may have formed.

The mode of drawing ploughs by locomotive engines, on the same plan that railroad carriages are drawn, does not offer a prospect of success, as the most powerful railroad engine would not even move itself upon an arable field, from the enormous weight of the machinery causing the wheels to sink into the ground; nor do we think that in the present state of mechanism it is possible to construct a steam engine of a sufficiently light weight to be allowed to go over the surface to be cultivated. Numerous attempts at running steam carriages on common roads have been made, and although some have for a time been loudly proclaimed as successful, they have all, one after the other, quietly made their exit: proving that however brilliant some of their performances may have been, they have not stood the unerring test of experience, and until the great talent that has been brought to bear upon this point has produced something permanently efficient, we believe it to be the wisest way not to attempt a matter beset with difficulties of greater magnitude, especially as there are so many profitable applications of this invaluable power that may be made upon farms of even moderate extent.

If ploughing by steam be ever accomplished, we think it will be carried out on the same or similar principles adopted in the steam plough of G. Heathcote, as carried out by that able engineer, Josiah Parkes, whose labours in the investigation of the principles of economy of steam engines are of the highest value: such full reports of this plough have been published that it will not be necessary to say more than that the plan adopted by him was to place the steam engine upon a carriage that could be moved along the headland of a large field, or a sort of road between two fields: two ploughs being drawn on each side of the engine if two fields were under operation at once, the ploughs being alternately drawn towards and from the engine by means of a band which passed round a pulley at the opposite end of the field. To carry out even this plan effectually, a farm would require to be laid out for the purpose; and although it may be brought to bear in large level tracts of country, where, from the extent of the work to be performed, and the capital at command, the prime cost of the machine may not be an obstacle, yet that in a densely peopled country like our own there are comparatively few localities to which it is applicable. We will therefore proceed to the consideration of those operations on farms that may be facilitated by the employment of stationary steam engines. When an intelligent agriculturist enters upon a farm for which he has to provide the whole of the stock, he carefully considers the locality, nature of the soil, and other matters, before determining upon the kind of horses he will employ, as well as the breed of cattle or sheep, and the implements required. The time is perhaps not very distant when a steam engine will be one of the matters to be thought of on every well conducted farm with 300 or 400 acres of arable land, perhaps even with less than this: it is therefore of some importance that attention should be paid to the rules for the selection of this kind of stock, as there certainly exists as great a difference in the quality and value of steam engines as there is between stock of the purest blood and those of the veriest dunghill breed that can be found; there being, however, this advantage on the side of the steam engine, that we can to a certainty have the best if we please to pay the price: we can understand all its disorders, and infallibly apply a cure; and when not required to work, it will cease to require feeding. Steam engines are usually classed under the two divisions of condensing and non-condensing: the former being very commonly called Boulton and Watts' engines, and the latter High Pressure. In no department of mechanics has there been a more numerous progeny of absurdities, and it will be a very desirable part of the education of farmers to be able to distinguish, in some degree, between improvements and mere alterations. This is not the place to discuss all the peculiarities of steam engines, and the proportionate economy of the various kinds actually in use, as the largest farm will require a much smaller engine than would suffice to carry out that extraordinary economy in fuel which the Cornish engineers have attained: it may, however, be desirable to give some general description of the steam engine best adapted to peculiar localities.

The condensing steam engine is so called from the circumstance of the steam, after having performed its office of driving the piston, being discharged into a vessel having a stream of cold water admitted that condenses it, and from whence it is pumped or discharged in the state of water. The non-condensing engine, on the other hand, discharges the steam direct into the atmosphere, and of this kind are all locomotive engines. These are more simple in construction and of course less costly to erect; but they consume considerably more fuel, as is ably shown by Mr. Parkes, in his elaborate papers, recently published in the Transactions of the Institution of Civil Engineers, and he estimates that a non-condensing engine working the steam at what is called 40lbs. per square inch requires just about double the fuel that a condensing engine takes. This estimate we have practical experience to corroborate. In all cases steam engines should be of workmanship of the best quality, and in all situations where coal is expensive this remark applies with increased force. A horse who could not do his work without a bushel of corn a-day would be dear at a gift, and an ill-constructed steam engine belongs to the same class of unprofitable articles. It is best to inclose it in a tight room so that the bright work may readily be kept clean, and the man who has the charge of it should have every inducement to exercise care and cleanliness, which are well repaid by increased durability. Wherever an abundant supply of water can be obtained, a condensing engine is to be preferred, from its greater economy of working. A well made four-horse power condensing engine may be estimated to consume about 50 lbs. weight of coal per hour, when at *full* work, and if only partly loaded the consumption may very probably not exceed 35lbs. per hour. The consumption of water will be about 14 gallons per minute, and where water is not abundant a pond of 100 square yards area, and 4 feet deep, will keep it supplied, as by cooling the water it may be worked an indefinite time, a small quantity suffices to supply the waste by evaporation. It will not be far from a right estimate to allow one horse power for each additional hundred acres of arable land upon the farms.* The more prominent purposes to which such an engine may be applied are thrashing and dressing corn, grinding corn for the use of stock, crushing oats, beans, oil cake, cutting

* The writer has had the opportunity of closely watching the working of a disc engine, of five horse power, both as regularly used in the factory and as applied to agricultural purposes. It continues to do its work admirably, and its consumption, when working up to five horse power, is about 14 lbs. of coke per horse power per hour.

roots, chaff cutting, churning, bruising gorse; grinding bread corn may be added, but except for flour, not requiring to have the offal thoroughly separated, this is not of much importance, owing to the difficulty and skill requisite to dress millstones so as to grind wheat properly.

The boiler of the engine will afford the requisite means for steaming potatoes, or any other food, for cattle in a cheap manner. And here I would venture to suggest the importance of a thorough practical investigation of the advantages to be derived from a well organized system of supplying fatting stock with cooked food.

Assuming that cooked potatoes, for instance, are better for stock than raw ones, it will not be difficult to show the great economy resulting from steaming them, by means of a pipe from the engine boiler over the process of boiling. If ten bushels of potatoes are required to be cooked at once by boiling, a vessel must be provided capable of holding about 140 gallons, and it will take about 40 gallons of water to fill the interstices and cover the potatoes; a fire is then lighted, and much time expended before the water even boils, and it must be kept in that state for a time proportionate to the size and quality of the potatoes. The water has then to be drawn off and thrown away, thus a large quantity of heat has been expended that is not made available. It may also be fairly estimated that the fuel consumed under vessels set for this purpose, as common kitchen or brewing coppers, does not make more than two-thirds the quantity of water boil that it would if applied under the steam engine boiler, so that here is an additional sourse of loss. There is also fetching the quantity of water to be taken into account, which, oft repeated, becomes of importance.

Let us now examine how the same effect is to be produced by means of steam taken from the boiler of the engine. The same vessel may be used for holding the potatoes, putting in merely a false bottom of metal or wood, pierced with holes. A pipe of one inch diameter is ample to convey the requisite quantity of steam, and which merely requires to be inserted beneath the false bottom; a close cover must be placed upon the vessel; some water will be condensed with the potatoes, which may be drawn off from time to time, the whole quantity may be a little more than that which is evaporated by the boiling process. It will be found that by this plan the ten bushels are cooked in one-half the time, with not more than one-half the fuel, probably not exceeding a third.

CONCLUSION.

Having thus gone through the general history and description of the various implements used in agriculture, I have only, in conclusion, to express my grateful acknowledgements to those agriculturists, and manufacturers of implements, who have so readily and so copiously furnished me with information. I have limited my remarks on the implements in modern use, to those which have fallen in some degree within the range of my own observation. I am sensible that many valuable inventions may have escaped my notice. Should further opportunity arise, it will be gratifying to me to extend my knowledge of these, and for any information that may tend to render the work more useful I shall be most grateful. It is only indeed by the enlightened and generous co-operation of all interested in this great branch of national enterprise, and source of national prosperity, that the true economy of agriculture can be ultimately obtained, and by which alone it will be able to maintain its just position throughout the length and breadth of the land.

APPENDIX.

Having, in the prosecution of this work, and in other instances had occasion frequently to refer to the records of patents, and having experienced considerable difficulty in the search for want of an analytical index, the writer believes the introduction of the following list of those patents which have immediate reference to agriculture and agricultural machinery will be generally interesting and especially valuable to such of his readers as may at any time be engaged in the improvement of the Implements of agriculture. In the arrangement of these, he has been materially aided by a professional friend, whose kind assistance he is glad to have the opportunity of thus acknowledging.

LIST OF PATENTS FOR MACHINERY AND PROCESSES, USED IN AGRICULTURE.

SECTION I.

PATENTS FOR DRAINING LAND.

2nd Jan., 1628.	Burrell, contrivances for draining marsh land.
12th March, 1663.	Wayne, R., engine for draining of levels.
19th Oct., 1797.	Watt's and Co., machine for draining.
4th Feb. 1800.	Lambert, draining plough.

8th June, 1809.	Dobito, under draining plough.
8th June, 1810.	Hickford, draining plough.
18th May, 1819.	Cowper, under draining plough.
22d June, 1819.	Jordan, water wheel for draining.
15th May, 1832.	Heathcoat, machine for draining and cultivating.
31st May, 1842.	Watson, R., improvements in draining land and in engineering works.

SECTION II.—MANURING.

1st May, 1633.	Mowell, C., for a composition for manuring land.
5th July, 1636.	Shawe, J., manuring and improving all sorts of ground.
28th April, 1640.	Chiver, R., new method of manuring ground.
8th Nov., 1721.	Piper and Tyndale, prepared chalk and sea water.
19th Feb., 1729.	Liverings, peculiar compound.
30th July, 1773.	Van Haake, manure for land.
28th Feb., 1795.	Dundonald, Earl, applying certain saline bodies.
9th Jan., 1802.	Estienne, dried and powdered night soil.
21st Oct., 1806.	Fletcher, prepared gypsum.
17th July, 1835.	Poittevin, powder for producing manure and disinfecting night soil.
2d Aug., 1837.	Rosser, F. R., improvements in preparing manure.
7th Oct., 1841.	Daniell, J. C., improvements in manufacture of manure.
23d May, 1842.	Lawes, J. B., improvements in manures.
23d May, 1842.	Murray, J., improved method of combining materials for manure.
10th Aug., 1842.	Albert, D. F., Improved manuring powder.

SECTION III.—CULTURE AND TILLAGE.

17th May, 1637.	Chiver, R., for methods of improving land and courses of husbandry.
18th June, 1799.	Hayes, culture and tillage.
23d Jan. 1806.	Berriman, machine for preparing land
13th Jan. 1808.	Earles, ditto tillage and dressing.
19th June, 1810.	Adams, machine for cultivating and Tilling land.
26th May, 1818.	Dyson, ditto ditto.
2d Aug. 1837.	Rosser, A. R. F., improvements in preparing manure, and in cultivating land.
14th Oct. 1837.	Vaux, Thomas, improvement in tilling and fertilizing land.
24th April, 1838.	Vaux, Thomas, improvement in tilling and fertilizing land.
8th Nov. 1838.	Winrow, J., improved apparatus for destroying weeds and insects.
15th Dec. 1838.	Vaux, Thomas, improvements in tilling and fertilizing land.
8th Sept. 1842.	Crosskill, improvements in Rolling and cutting land used in culture.
7th July, 1841.	Vavasour, improvements in machinery for tilling land.
7th July, 1842.	Hall, J., improvements in ditto.

SECTION IV.—PLOUGHING.

25th Nov. 1623.	Hamilton, improvement for ploughing, gardening, and sowing arable grounds.
6th Aug. 1627.	Brouncker, V., machine for ploughing land without horses.

17th July, 1634. Perham, W. V., ditto.
21st Sept. 1730. Stanforth and Foltfamle, improved plough.
15th June, 1770. Moore, improved plough.
18th March, 2785. Ransome, plough shares.
13th Jan. 1783. Cooke, machine for ploughing and drilling.
12th Aug. 1788. Cooke, ditto.
16th Aug. 1791. Merricks, constructing ploughs.
26th April, 1792. Smart, plough shares.
25th May, 1798. Sanxter, Ploughs for preparing land.
11th Oct. 1800. Plenty, new invented plough.
6th Feb. 1802. Somerville, Lord, double furrowed plough.
30th Oct. 1802. How, improved plough.
24th Sept. 1803. Ransome, tempering cast iron shares.
30th May, 1808. Ransome, Swing ploughs.
30th May, 1809. Manley, plough for saving labour.
8th Oct. 1810. Hazeldine, improvements in ploughs.
26th March, 1811. Hazeldine, plough improved.
15th June, 1813. Cooke, ditto.
23d Sept. 1813. Liston, ditto.
5th Oct. 1814. Phillips, improved plough.
14th June, 1815. Brown, swings of Plough.
23d Aug. 1815. Bemman improved plough.
22d Dec. 1815. Plenty, ditto.
23d March 1816. Brown, improved swing plough.
1st June, 1816. Ransome, certain improvements in ploughs.
19th April, 1817. Nicholas, improved plough for covering over seed.
5th July, 1817. Wedlake, plough improved.
5th Aug. 1817. M'Carthy, ditto.
30th Jan. 1819. Thomas, ditto, and a propeller for
28th Nov. 1820. Ransome, ditto, improved.
1st May, 1821. Thomas, cutting up ground for tillage.
5th July, 1823. Clymer, plough for land.

15th Jan. 1824. Finlayson, plough and harrow, self-acting.
4th April, 1827. Stothert, improved plough.
1st July, 1830. Clive, ploughs locomotive.
19th July, 1832. Wedlake, improved plough.
2d Nov. 1835. Springall and Ransome, certain parts of ploughs.
28th Aug. 1837. Armstrong, W., improvements in ploughs.
4th Nov. 1837. Upton, J., steam plough harrowing &c.
17th June, 1839. Campbell and another, improvement in ploughs.
18th March, 1839. Campbell and another, improvement in ploughs.
24th Dec. 1839. M'Rae, A., machinery for ploughing, harrowing, &c.
25th Feb. 1840. Huckvale, improvements in ploughs.
25th March, 1840. Hay, J., improved plough.
16th April, 1840. Cooper, R., improvements in ploughs.
28th May, 1840. Campbell, improvements in ploughs.
18th July, 1840. Palmer, W., improvements in ploughs.
3d Aug, 1840. Sanders and others, improvements in ploughs.
15th Feb. 1841. Smith, T., improvements in ploughs.
9th May, 1842. Warren, J., improvements in ploughs.
10th June, 1841. Bentall, improvements in ploughs.
22d Sept. 1842. Sanders and others, improvements in ploughing and harrowing.

SECTION V.—DIGGING.

28th Jan. 1784. Van Thornhoff, machine for digging up ground.
20th Aug. 1823. Elwell, Improved spades and shovels.
27th March, 1839. Newton, W., digging and removing earth.

SECTION VI.—DRILL AND HOE PLOUGHS.

22d Aug. 1781. Proud, drill to be fixed to a plough for sowing.

18th Oct. 1786. Winter, drilling seed.

3d July, 1787. Wright, drill plough for sowing grain.

20th June, 1789. Ridge, drill and hoe ploughs.

3d Nov. 1801. Jackson, drill machine for sowing turnips.

27th July, 1813. Madeley, machine for drilling corn and turnips.

1st Nov. 1820. Torey, drill fixed to a plough for sowing.

19th May, 1827. Coggin, improvements in drilling grain.

3d Nov. 1838. Baron Western, C. C. improvements in drills for corn, grain, seeds, &c.

11th Jan. 1839. Newton, improvements in machinery for drilling land, or sowing grain of every description.

23d April, 1839. Miller, improved drilling machines.

12th June, 1839. Grannsel, improvements for drilling corn, grain, pulse, &c.

25th Nov. 1839. Hornsby, improved drilling machines.

25th Feb. 1840. Huckvale, improvements in ploughs.

31st Dec. 1840. Hensman, improvements in ploughs.

15th Feb. 1841. Smith, improvements in Ploughs.

13th June, 1842. Garrett's, R., improvements in horse-hoes, drag-rakes, and drills.

SECTION VII.—SOWING.

2d Jan., 1634. Ramsey, D., method of sowing corn and grain.

18th Feb., 1639. Platt, G., Setting corn and meliorating barren lands.

13th March, 1784. Horn, a machine for sowing seed.

30th July, 1784. Wright, a machine for sowing corn.

20th Oct., 1785. Horn, a machine to be fixed to a plough for sowing.

29th Oct., 1788.	Hele, a machine for sowing grain.
27th Oct., 1789.	Boorn, a machine for sowing corn.
19th Aug., 1790.	Perkins, ditto ditto.
26th April, 1800.	Richards, a machine for setting and depositing grain and seeds.
17th April, 1806.	Plucknett, machine for dibbling and drilling seed.
4th Dec., 1813.	Tyrrel, a machine for sowing corn.
19th May, 1827.	Coggin, T. P., a machine for dibbling grain.
2d Nov., 1835.	Keene, machinery for sowing corn.
2d Dec. 1839.	Saunders and Newberry, machinery for dibbling, or setting wheat and other grain.
12th May, 1840.	Bradshaw, improvements in dibbling corn and seed.
22d May, 1840.	Rham, improvements in machinery for preparing land and sowing seed.
8th Aug. 1840.	Edwards, improvements for preparing and drilling.
19th June, 1841.	Shaw, improvements in setting wheat and other seeds.
7th June, 1842.	Irving, improved corn drill for sowing all kinds of grain.

SECTION VIII.—HARROWING.

(See also section vi.)

10th March, 1787.	Heaton, drill harrow for sowing and harrowing simultaneously.
10th March, 1798.	Lester, harrow of a new construction.
17th Feb. 1801.	Wilde, improvements on harrows.
6th June, 1833.	Madely, improved harrow.
13th June, 1836.	Vaux, agricultural harrow.

21st April, 1838. Finlayson, improvements in harrows.

30th May, 1839. Armstrong, improvements in harrows.

22d Sept. 1842. Sanders, improvements in ploughing and harrowing land, and in cutting food for animals.

SECTION IX.—REAPING.

(See also mowing, section x.)

4th July, 1799. Boyce, machinery for cutting wheat and other grain.

20th May, 1800. Meares, ditto ditto.

31st July, 1809. Hutton, improved reaping hooks.

26th July, 1811. Cumming, reaping machinery.

21st Sept. 1814. Dobbs, machinery for gathering and cutting corn.

20th May, 1842. Phillips, improvements in reaping and in cutting food for cattle.

SECTION X.—MOWING.)

(See also reaping, section ix.

30th June, 1788. Sandilands, machine for cutting grass.

17th Dec. 1791. Hill, improved coustruction of scythes.

15th June, 1805. Plucknett, mowing machinery.

23d Aug. 1805. Ditto, ditto.

26th Aug. 1807. Hill, iron and steel backs to scythes.

26th Aug. 1813. Hunt, improved backs to scythes.

31st Aug. 1830. Budding, machine for mowing grass, lawns, &c.

2d Nov. 1840. Duncan, improvements in cutting and reaping grass, &c.

SECTION XI.—THRASHING AND SEPARATING MACHINES.

(SEE ALSO SECTION XII.)

13th Feb. 1734.	Menzies, machine for thrashing grain.
21st June, 1785.	Winlaw, mill for separating grain from straw.
9th April, 1788.	Meikle, A., ditto ditto.
21st Feb. 1792.	Willoughby, machine for thrashing corn.
19th Feb. 1795.	Jubb, machine for thrashing and winnowing corn.
2d June, 1795.	Wigfull, machine for separating the grain from straw.
31st Oct. 1796.	Steedman, thrashing machinery.
4th July, 1797.	Maule, machine for cleaning the grain from straw.
5th June, 1798.	Palmer, apparatus for cleaning grain from straw.
9th Nov. 1799.	Tunstall, machine for thrashing all kinds of grain.
6th Dec. 1799.	Palmer, machine for cleaning grain, and cutting straw for cattle.
19th June, 1802.	Lester, machine for separating the grain from straw.
18th May, 1804.	Burrell, thrashing machines.
30th Oct. 1804.	Noon, thrashing machines with loose beaters.
16th Jan. 1805.	Lester, engine for separating corn, &c., from the straw.
5th Feb. 1805.	Ball, thrashing machine.
ditto.	Perkins, thrashing machines.
23d Nov. 1805.	Lambert, improved thrashing machinery and a windlass.

21st Nov. 1807. Lester, machine for separating corn from straw.

31st Oct. 1808. Andrews, thrashing machine.

23d Jan. 1810. Cox, ditto.

22d May, 1810. Onions, ditto.

29th June, 1813. Todd, separating grain from the straw.

27th Sept. 1814. Lister, machine for separating the grain from straw.

3d Dec. 1817. Wild, machine for separating corn from straw.

7th May, 1840. Atkinson, improved thrashing machine.

1st Oct. 1840. Mackelcan, F., improved thrashing machine.

21st Jan. 1841. Cooper, ditto.

28th Jan. 1341. Pryor, ditto.

2d Aug. 1842. Dry, J., improvements in thrashing machines.

SECTION XII.—WINNOWING AND DRESSING.

17th June, 1800. Cooch, winnowing machine.

16th Dec. 1811. Elvey, ditto.

23d March, 1839. Salter, F., dressing corn and other grain.

5th Feb. 1818. Smith, winnowing machine.

SECTION XIII.—HAY MAKING AND CUTTING.

22d Aug. 1814. Salmon, machine for making hay.

27th July 1816. Ditto, ditto.

18th Dec. 1793. Hill, hay knives for cutting hay.

SECTION XIV.—CHAFF CUTTING.

2d June, 1794.	Naylor, machine for cutting chaff.
8th Jan. 1794.	Cooke, ditto.
4th Feb. 1800.	Lester, machine for cutting hay and straw.
17th Feb. 1800.	Lester, chaff cutting machine.
23d July, 1802.	Lawden, machine for cutting straw.
7th Feb. 1804.	Passmore, machine for chopping straw.
4th Feb. 1808.	Snowden, chaff cutting machine.
14th June, 1815.	Gardner, machine for cutting hay.
7th March, 1818.	Heppenstall, machine for cutting straw.
1st Nov. 1819.	Shorthouse, ditto.
6th July, 1840.	May, C., improvements for cutting straw, hay, &c.
15th Oct. 1840.	Lord Ducie and others, machinery for cutting vegetable substances.
27th Oct. 1842.	Gardner, James, implements for cutting hay, straw, and other matters for animals.

SECTION XV.—CHURNING.

8th Aug., 1777.	Rastrick, barrell churn.
28th July, 1792.	March, horizontal churn.
10th Nov., 1796.	Raley, butter churn.
9th May, 1807.	Wood's churning machine.
17th May, 1842.	Williams, T., improved churn.

SECTION XVI.—OPERATIONS, CHIEFLY HORTICULTURAL.

22d Jan., 1624.	Shipman, sowing, setting, and planting madder.

14th April, 1637. Everadd, for ordering and contracting of saffron and other vegetable.

3d Feb., 1671. Burneby, invention for ordering of rice and safflower.

16th March, 1724. Greeing, grafting or badding English elm upon Dutch.

5th Dec., 1785. Le Brocgr, rearing and cultivating fruits.

17th March, 1808. Weeks, forcing frame for raising cucumber and other fruit.

7th Nov., 1812. Jukes, shears for pruning trees.

28th Feb., 1824. Richards, metal frame for hot houses.

9th Jan., 1828. Grubble, heated walls for ripening fruits.

13th Aug., 1830. Knowles, hop pole drawer.

6th Oct., 1830. Harrison, improvement in horticultural buildings.

2d Nov., 1839. Catten, improvement in garden pots.

SECTION XVII.—MISCELLANEOUS IMPROVEMENTS, OPERATIONS, AND PROCESSES.

1st June, 1731. Riley and Beaumont, composition to feed swine.

11th Oct., 1797. Baker, method of preventing the smut in wheat.

21st June, 1803. Brown, cutting turnips, carrots, &c., for animals.

20th April, 1809. Wilks, compound cake for feeding horses and other animals.

26th Feb., 1810. Pratt, machines for preparing various agricultural operations.

20th Aug., 1811. Jorden, improved mode of glazing hot houses.

30th Oct., 1811. Martyn, instruments for hoeing turnips.

7th April, 1813. Timmins, improvement in hot houses and other horticultural buildings.

26th May, 1836.	Blurton, extracting milk from cows.
26th May, 1814.	Neville, hurdle and gate making.
7th June, 1814.	Jorden, improvement upon horticultural building.
23d April, 1825,	Roberts, T. A., preserving potatoes and other vegetables.
12th Aug., 1830.	Knowles, improved hoppole drawer.
6th Oct., 1830.	Harrison, improvement in glazing roofs, &c.
22d March, 1832.	Young, mangel wurzel for producing articles of commerce.
7th March, 1833.	Springall, corn stack stand.
25th Sept., 1834.	Gardner, machine for cutting turnips.
11th Jan., 1837.	Gardner, improvement for cutting turnips, &c., for cattle.
8th Nov., 1838.	Winrow, improved apparatus for destroying weeds and insects.
12th Dec., 1838.	Gardner, improvements in cutting turnips for cattle.
7th Nov., 1839.	Moody, E., preparing roots for animals.
9th July, 1840.	Payne, preserving vegetable matters.
29th July, 1840.	Worth, cutting vegetable substances.
15th Oct., 1840.	Francis and other, improvement in cutting vegetable substances.
25th Nov., 1840.	Grellet, modes of treating and converting articles of food.
8th Sept., 1841.	Grant, improved horse rakes and hoes.
8th Sept., 1841.	Croskill, machinery for rolling and cutting land.
26th Sept., 1841.	Huckvale, improvements in horse hoes for treating and dressing turnip.
29th Sept., 1841.	White, improvements in horse hoes.
7th March, 1842.	Green, improvement in machinery for cutting turnip for cattle.
31st March, 1842.	Daniell, improvement in making and preparing food for cattle.

SECTION XVIII.—CLEANSING AND DRYING GRAIN.

25th Nov., 1715. Masters, cleansing and curing the Indian corn.

26th April, 1725. Woodruffe, engine for washing corn.

10th Sept., 1789. Hancock,. machine for cleansing and shifting grain.

15th April, 1802. Gardner, preserving and purifying damaged grain.

6th July, 1803. Roberts, clearing smut from wheat.

23d May, 1826. Hughes, restoring foul or smutty wheat.

6th July, 1830. Tuxford, cleansing and purifying wheat.

12th Dec., 1830. Barlow, cleansing wheat.

24th Jan., 1833. Hebert, washing, scouring, separating, and drying machinery.

2d June, 1836. Hebert, kiln for drying wheat, &c.

7th June, 1836. Berry, machinery for cleaning and drying wheat.

3d Dec., 1836. Don, Thos., improvement in drying and preparing grain, seeds, berries, and in manufacturing their products.

11th Jan., 1838. Jontham, machine for drying corn and other grain and seeds.

3d Nov., 1838. Hebert, Luke, apparatus for storing, cleansing, and preserving grain.

15th April, 1840. B. De Los Valles, preserving corn and other grain.

11th Jan., 1841. Newton, W., machinery for cleansing wheat, &c., from smut, &c.

14th Jan., 1841. Hall, seed and dust disperser for freeing corn, &c., from insects.

14th March, 1768. Meikle and Macknell, dressing and cleansing wheat from smut.

SECTION XIX.—MILLS FOR GRINDING CORN.

8th March, 1716. Thompson, floating engine for grinding and bolting.
6th Sept., 1744. Perkins, machine for grinding corn.
24th Nov., 1750. Perkins, machine for grinding corn.
24th July, 1752. Wilkinson, metal rolls for crushing grain, &c.
24th Dec., 1767. Hayne, mill for grinding corn.
6th Oct., 1768. Freeth, hand corn mill.
30th Oct., 1768. Taylor, machine for grinding corn.
26th Nov., 1774. Watson, handmill for grinding corn.
11th June, 1775. Rawlinson, mills for grinding and cleansing beans, &c.
23d March, 1779. Dearman, mills for grinding corn.
23d Aug., 1780. Pickard, mills for grinding corn.
21st June, 1783. Winlaw, mill for grinding corn.
8th Dec., 1785. Maunsell, horizontal windmill for corn.
24th Jan., 1797. Ferryman, blanching, grinding, and drying corn.
2d Aug., 1799. Hunt, machine for grinding corn with steel and iron plates.
25th April, 1800. Wright, hand stone corn mill.
2d Aug., 1810. Williams, machine for grinding malt and splitting beans.
28th Aug., 1801. Hawkins, floating corn mills.
18th Sept., 1801. Barrett, corn mill.
28th Jan., 1812. Taylor, reducing grain into flour by machinery.
1st April, 1814. Smart, machinery for grinding corn.
23d May, 1815. Kenrick, mills for grinding coffee, malt, &c.
11th Sept., 1823. Hase, mills for prisons.
8th Jan., 1824. Devereaugh, mills for grinding corn.
2d Nov., 1829. M'Curdy, mills for grinding corn and composition for stone.

15th Dec., 1831. Savoye, metallic corn mills, vibrating motion.

24th Jan., 1833. L. Hebert, progressive corn mills and dresser.

7th March, 1833. Don, corn mills, revolving in vertical planes.

1st Jan., 1834. Sharp and Roberts, corn mill, with eccentric runners.

13th Sept., 1834. Berry, corn mill, revolving in vertical planes.

10th Aug., 1835. Hebert, combination of mill and dressing apparatus, &c.

2d June, 1136. Hebert, new combinations and improvements.

19th March, 1831. Horsfield, Wm., improvement in construction of mills for grinding corn.

24th Sept., 1840. Dean and Evans, mills for grinding corn.

13th Sept., 1841. Scott, improvement in flour mills.

21st March, 1842. Parkes, Z., improvement in machines for grinding and dressing wheat, &c.

19th Jan., 1843. Hebert, L., improvement in machinery for grinding and dressing grain.

SECTION XX.—MILL STONES.

1st March, 1796. Pratt, composition for artificial mill stones.

27th April, 1824. Dallas, machinery for dressing stones.

15th Sept. 1829. Milne, dressing stones.

15th Dec. 1842. Poole, M., improvements in dressing mill stones.

SECTION XXI.—SIFTING AND BRUSHING MEAL.

21st July 1623. Rathbone and another, machine for bolting the meal.

23d June, 1686. Finch and another, woven wire for bakers, mealers, and others.

10th May, 1765. Milne, wire cylinders for dressing flour.

23d Oct. 1770.	Milne, wire cylinders for dressing wheat
24th Dec. 1781.	Smith, dressing and bolting flour.
19th Dec. 1783.	Blackmore, bolting cloths for dressing flour.
21st March, 1800.	Blackmore, improvements in making bolting cloths without seams.
23d March, 1804.	Bowntree, machinery for agitating and separating.
19th Feb. 1825.	Ayton, springs to bolting machines.
4th June, 1829.	Smith, dressing flour.
3d May, 1832.	Goodlet, preparing rough meal for ground corn.

SECTION XXII.—MALT.

28th Nov. 1698.	Groves and Reeves, malt upon turned iron.
1st Dec. 1713.	Bird, engine for drying malt and hops.
23d Jan. 1720.	Desagutiers, using steam in drying malt, hops, &c.
17th Nov. 1743.	Southgate, curing malt upon metal floors.
21st Feb. 1769.	Wildlay, improvements in drying malt.
15th April, 1778.	Thorntons, solid extracts of malt, hops, &c.
14th Nov. 1783.	Clark, improved malt and oat kiln.
15th May, 1792.	Whitmore, machine for making malt.
9th Jan. 1796.	Pepper, kiln for drying malt.
8th May, 1798.	W. Jones, machine for mixing malt.
29th Jan. 1805.	Barrett, improvements in malt kilns.
19th Dec. 1807.	Salter, apparatus for drying malt and hops.
2d Aug. 1810.	Williams, machine for grinding malt.
26th Nov. 1811.	Adams, improvements in preparing malt.
1st Nov. 1816.	Varley and another, producing saccharine from barley, &c.

28th March, 1817.	Wheeler, improvements in making malt.
10th June, 1817.	Whittle, kiln for drying malt.
5th May, 1818.	Bush, ditto, ditto.
1st June, 1819.	Geldart, heating kilns for various purposes.
8th July, 1829.	Salmon, improvements in malt kilns.
10th Feb. 1825.	Lambe, composition for malt and hops.
20th June, 1820.	Vallance, packing and preserving hops.
11th Nov. 1830.	Pratt, perforated quarries to malt kiln.
7th Sept. 1833.	Else, kiln for preparing malt.
10th July, 1834.	Page, method of drying malt with sea coal, peat, &c.
24th Aug. 1837.	Boulay and another, improvements in drying malt and screening.
24th Aug. 1837.	Brown, improvements in cockles stoves for drying malt.
22d Sept. 1842.	Stead, improvements in the manufacture of malt.

INDEX.